Cyber-Archaeology

Edited by

Maurizio Forte

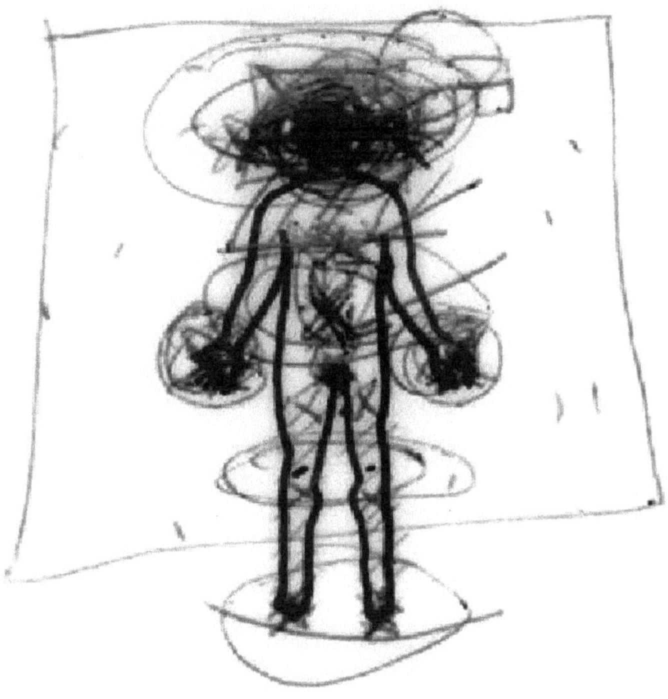

BAR International Series 2177
2010

Published in 2016 by
BAR Publishing, Oxford

BAR International Series 2177

Cyber-Archaeology

ISBN 978 1 4073 0721 3

BAR Publishing is the trading name of British Archaeological Reports (Oxford) Ltd. British Archaeological Reports was first incorporated in 1974 to publish the BAR Series, International and British. In 1992 Hadrian Books Ltd became part of the BAR group. This volume was originally published by Archaeopress in conjunction with British Archaeological Reports (Oxford) Ltd / Hadrian Books Ltd, the Series principal publisher, in 2010. This present volume is published by BAR Publishing, 2016.

Printed in England

BAR
PUBLISHING

BAR titles are available from:

BAR Publishing
122 Banbury Rd, Oxford, OX2 7BP, UK
EMAIL info@barpublishing.com
PHONE +44 (0)1865 310431
FAX +44 (0)1865 316916
www.barpublishing.com

Contents

From Archaeology to I-archaeology: Cyberarchaeology, paradigms, and the end of the twentieth century

Ezra B.W. Zubrow
Department of Anthropology
University at Buffalo
Yip Fellow Magdalene College

I am told that some historians believe in what are termed the "long centuries". So the seventeenth century ends in 1715. The eighteenth century of the European and American Revolution finishes in 1815; the nineteenth century terminates in 1914 with the assassination of the Grand Duke in Sarajevo and the plunging of the world into war. It is thus appropriate in 2011 to reconsider the "long" previous century and try to prognosticate what will happen in the next.

Thomas Kuhn suggested that there were paradigm shifts in science. They occur when the fundamental theories, hypotheses, data and underlying assumptions are questioned and demonstrated to be wrong and new fundamental concepts replace them. Sometimes one will occur during a century. Occasionally, they will be more frequent and sometimes less. However, the reality is that there are not only paradigm shifts, there are far broader shifts that take place cross cutting not only many academic fields simultaneously, but, the way people think and the way they live. It is not surprising that the "long" centuries occur. By fifteen or twenty years into each century, those people born in the last decades of the previous century are young or mature adults and are influential enough to reject the intellectual and cultural heritage of the previous period. They believe the former belongs to their parents and clearly is not their own. These "new" thinkers, intellectuals, cultural trend setters, politicians, bureaucrats, etc. will with their children live the rest of their lives in the "new" century.

If a century is to have a character, it must be established by the first twenty years of its existence and the event that characterizes it should have occurred. What is it for the 21 century? (And how will it apply to archaeology? Is cyber archaeology a change in methodology, a change in paradigm, or a reflection of a broader change?)

Perhaps a brief word about the twentieth century is appropriate. It was the century of globalization. Beginning with large world encircling empires, two global wars in the first half of the century, the rise and worldwide spread of the internal combustion engine in trucks and cars, mail, radio, television, computers, internet and global markets were complemented by the rise of global organizations. By the end of the century markets rose and fell on the internet, persons were met and married by pc technology and what took place in Japan in the early morning 3:00 am was old news in NY by the afternoon 2:05 pm five minutes later. Capital is global even if labor was not. The defining event surely was World War I.

For archaeology the twentieth century was similarly characterized by globalization. It is characterized by the outlining of the world's prehistory and then the filling in the geographic gaps and refining the chronologies through new and better dating technique(Clark 1961; Clark 1969). By the end of the century it would be fair to say that for most areas of the world the broad outlines of world prehistory are known and that most work now takes place within that framework. Of course, there always are new discoveries that occasionally change the details or even the basic outline but these are rare and usually are not too important changes when taken in the global perspective of world prehistory.

The other two major changes have been the "scientification" of archaeology through new scientific techniques and the "new archaeology" (Clarke 1968; Binford and Quimby 1972; Binford 1977) on the one hand and the "interpretationist" views of post processual archaeology (Hodder 1986; Shanks and Tilley 1987; Hodder 1989; Shanks and Tilley 1992; Tilley 1993; Tilley 1999; Hodder 2001). Although some movements and ideas have fizzled others have tended to continue in their original or modified forms. Thus, cultural history still plays a major role in the everyday life of the everyday archaeologist around the world. So do many of the scientific aspects of methodology of the "processual or new archaeology". Techniques of discovery, (remote sensing including electrical resistivity, side looking and ground penetrating radar, seismography etc.), environmental reconstruction (palynology, isotope analysis, dendroclimatology, diatom analysis, zooarchaeology, etc.), dating (radiocarbon, archaeomagnetism, thermoluminscence, etc.), data handling (sampling, quality control, databases, metadata, data mining, data merging, etc.), landscape location and reconstruction (GIS, GPS, Satellite imagery, etc.), and many other scientific techniques have become part of the standard tool kit of the archaeologist throughout the world. They are applied to archaeological problems in a "scientific context". Not all are applicable to all problems or areas, nor are all archaeologists able to afford all the scientific techniques but most are aware and would use them if possible.

It is less clear that the "interpretational" school in its pure forms will continue. With the general decline in "post modernism", the inability of the post processual researchers to build upon their many important insights made in the twentieth century, their rejection of machine intelligence research and cybernetics, and, their inability to relate to many important discoveries in other disciplines, their future is unknown. Students speak of "processualism plus" never "post processualism plus" or for that matter "minus". Rather as this volume of cyberarchaeology shows clearly the most substantial, applied, and useful aspects of the post-processual thought are in cyberarchaeology. If it is to continue it will continue using the theoretical underpinnings of cybernetics. In other words, today one might say post processual archaeology must be cyberarchaeology if it is anything.

However, as this book shows both processual and post processual are now integrated into something new. Although reductionist investigations still form an important component of present-day archaeology, they are not the only way to make progress. As far back as Aristotle there has been another view- a more holistic interpretationst view. Namely, the sum is greater than the whole of its parts and as one moved through economic, environmental, systems and post processual archaeology clearly this view becomes the most important alternate relevant direction. And as more and more people examine the relevance of the past -archaeologically and historically -for the environmental and ethnic aspects of the modern world, this view becomes more relevant. Instead, a new language of archaeology develops in which complexity, organization, uniqueness, emergence, holism, unpredictability, openness, interconnectedness, teleonomy, disequilibrium, immersion, and evolution became more dominant terms."

So in a very real way, cyber archaeology bridges the gap between "scientific" and "interpretational" archaeology for it provides testable in the sense of adequacy material representations of either "interpretations" or "scientific hypotheses or discoveries."

If one tracks the global spread of archaeology, it seems as if theory has the most rapid spread rate, then in descending order methodology and data. But the order seems to be the same for the decay rate with theoretical ideas being abandoned most rapidly, then methodologies, and finally substantive data.

Archaeology is in many ways behind the global curve for the development of ideas. A major prerequisite for globalization is standardization. Most archaeology is under regional and national standards. A search of the ISO, the International Organization for Standardization, under archaeology brings up only one document –"Information and Documentation-A reference ontology for the interchange of cultural heritage information" (Standardization 2006). Useful but it is not archaeology. One applies to the local, municipal, regional, provincial or national office for a permit to survey or excavate. The standards for each of these both within and among countries differ widely. There has been some limited inroads created by international treaties such as the Unesco Convention on the Means of Prohibiting and Preventing the Illicit Import, Export and Transfer of Ownership of Cultural Property 1970 regarding definitions and expected legal behavior(O'Keefe and Institute of Art and Law (Great Britain) 2007).

Cybernetics is defined by the Oxford English Dictionary as "Of, relating to, characteristic of, or involved in (the culture of) computers, information technology, and virtual reality; futuristic." It is used broadly in our language and there is a connotation that cyber is used for computer or electronic counterparts of a pre-existing product, service or phenomena. Originally prefixed to nouns it was limited to information technology, computers, and virtual reality. But it has extended to the Internet and well beyond. In addition to the topic of this book-cyberarchaeology, there are cybercultures, cybernations, cybraries, cybercubicles, cybernauts, cyborgs, cyberfeminists, cyberlovers, cybercash, cybercops, cybermail. All of us, at one time or another, have used cyberspeak as well as having coffee at a cybercafé. With a more negative cast, there are cybercrime, cyberporn, cybercide, cyberphobe, cyberpunk, cyberchrondiac, cyberterrorism, and cyberwars within the broadly construed cyberworld.

The word is derived from the Greek steersman "Kubernetes". "Cyber" has the connotation of "control" or "being controlled". A connotation that virtual does not have. In some circles it has been replaced with "e" for electronic as in "e-democracy", "e-banking", or as we all use "e-mail" now "email". In 1947 Turing published his paper "On Intelligent Machinery" in which he argued that natural intelligence is to be examined in terms of science(Turing and Copeland 2004). It was quickly followed by Weiner's book published by MIT in 1948 entitled Cybernetics or Control and Communication in the Animal and the Machine(Wiener 1948). It was the first to use the term. He examined the similarities of control between biological and mechanical systems and focused on the importance of self-controlling systems and regulators. Cyber-archaeology continues to do so. In 1958, there was a London conference "On the Mechanization of Thought Processes" that laid the framework for "machine intelligence". It took Turing's and Weiner's major innovation to another stage asking "what is the difference between "self regulation" and "thinking"? Can machines think and if not what do we need to do to make them think"? Archaeology is beginning to ask about the importance of self-thinking machinery for survey, for excavation, and most frequently for exhibition and communication of results for the general public.

The founding fathers and mothers of cybernetics" included Frank Rosenblatt (inventor of the Perceptron (Rosenblatt 1962); Arthur Samuel (inventor of the first learning algorithm)(Samuel 1995); Marvin Minsky (one of the founding fathers of AI)(Minsky 1963); Oliver Selfridge (inventor of the Pandemonium architecture, a paradigm for modern agent systems)(Selfridge, Rissland et al. 1984); John McCarthy (inventor of LISP, and of the name Artificial Intelligence)(McCarthy, Massachusetts Institute of Technology. Artificial Intelligence Laboratory. et al. 1959; Massachusetts Institute of Technology. Computation Center., McCarthy et al. 1960); Donald MacKay (a biological cyberneticist) (MacKay 1954) Warren McCulloch (co-inventor of the neural networks model still used today)(McCulloch and Pitts 1943); Ross Ashby (inventor of the concept of homeostasis)(Ashby 1956); Grey Walter (roboticist)." Many archaeologists use either directly or indirectly "artificial intelligence" with their "expert systems", "agent simulations", and their "neural networks".

A long time ago in cybernetic but not archaeological terms an algorithm was introduced called Viterbi's algorithm (Viterbi and Omura 1979) (Viterbi 1995). It was shown in the 1960's that there was a way to determine what was the most probably correct message when a corrupt message was received. Not only did it mean a big contribution on how information was to be processed but also it helped change the criteria of success. One did not need to reach the right answer every time nor represent a phenomena in a computer in exactly the same way as it was originally formulated.

Cyber became equivalent to computing in the 1970s as my colleagues and I were using the Cyber computers of Control Data Corporations (CDC). There was the popular Cyber 170 and later the Cyber 200 series that moved to byte architecture and time-sharing. Designed by Seymour Cray (the primary architect of the supercomputer) the CDC Cyber computers were among the best in the world but eventually died with the PC and CDC filed bankruptcy in 1995.

A major step in the 1980's in the development of cybernetics was to stop trying to simulate the way humans would do or represent a task but to emulate the desired behavior. It put machine learning as one of the central aspects of cybernetics and one has DDAI or data driven artificial intelligence. The relationship of people to machines became a central in a wide range of work including the seminal work of Terry Winograd (Winograd 1972; Winograd 1983; Winograd and Flores 1986) that recognized that it was not how the communication between people and machine took place but that it did take place –there were many ways to listen and then others recognized there were many ways to see. His student Larry Page founded with Sergey Brin a company called Google. Similarly, Jack Dangermond at the Harvard Graphic Laboratories in the 1960's began to wonder about computerized mapping and over the next decades helped lead the Geographic Information System revolution that not only created new spatial algebras and statistics as well as techniques to revisualize and relocate places in the world but became a multi billion dollar industry touching business, media, defense, education, government, agriculture, environment, mining, oil, emergency relief, aviation, public utilities, and health to name a few. Archaeologists took to GIS with a vengeance and it is now in every archaeologists tool kit(Allen, Green et al. 1990).

With increased power of computing the node of analysis began to change from the population to the individual and 'agent' modeling and simulation spread from the population in the hard sciences to the individual in the social sciences such as economics, sociology and archaeology. To recognize how far this has moved, one need only note the Sante Fe Institute –founded by was founded in 1984 by a set of Los Alamos scientists including Cowan, Colgate, Metropolois, Anderson, Carruthers and Slansky as well as Gell-Mann and Pines. It is now headed by the archaeologist Jeremy Sabloff and includes the archaeologists Sandor Van Der Leeuw, Linda Cordell and George Gummerman.

So what happened is there developed in the late twentieth century a new criteria for scientific and even more broadly visual and sound representations of reality. Rejecting the notion of problem solving as being only way to do science, has freed the disciplines. Computers do not have to understand "how to create an atomic explosion atom by atom," "how to show an object", "how to find sites", or" how to reconstruct Petra exactly stone by stone". Instead by analyzing thousands of digitized examples, they are able to emulate the process "well enough". It is a similar to a language student. One may pass the grammar or vocabulary tests similar to a scientific test. Or one may speak with a native and if the native understands the student then the student has "succeeded well enough" even if the grammar or the vocabulary might need improvement.

Individual understanding is the ultimate criteria for success. In fact the rules the student (the computer uses) and the native (science) may be very different. What is important is that they get to the same place. This also allows for one's ability to now model intelligent agents to pursue their goals in complicated environments and make informed decisions about what is perceived, known and remembered. One need only note how close Amazon can come to making a recommendation on what one should read or e-bay on verifying the business trustworthiness or their sellers. They and Google are autonomous web systems whose explicit reasoning may not be explicitly logical or one hundred percent completely applied to the domain but they do it "well enough" that it actually works.

Archaeologists today are concerned with improving the observers' intuitive understanding of past events and memory (Costopoulos 2001) Given the above, one should take as given that all of the work in cyberarchaeology represents the best of both scientific and interpretationist archaeology. In fact one could argue that if post-processual archaeology will continue to exist it will exist through cyber archaeology. It is in cyberarchaeology where the interesting issues of cognition, memory, individual difference, education etc are actually being researched and actually being used. Each of the articles in this volume sit at different places on the this processual post processual continuum. They also sit at different places on other axes as well. Since one is unable to present a nine dimensional figure in a textual format, one must do with three three dimensional diagrams. However the axes are the continua from processual to post=processual, private to public, research to exhibition, cognitive to behavioral, theory to methodology, methodology to data, textual to visual, innovation to review, and historical to futuristic.

(See figure 1).

What is determined to be "good archaeology" for cyberarchaeology is not just the discovery of new sites, new methodologies, new cultures, new processes, or even new interpretations about the past. It is not only about communication. As these articles show, there are much broader criteria.

(See table 1)

So what is the event that will mark and characterize the next century and what will be the death knell of the twentieth. One might look to world politics and see the US as the sole super power, or the 9 -11 attack, or, the reality of terrorism. Or one could turn to ideology and suggest that the Iraq Afghanistan wars are the beginning of the Orwellian vision of three ideologically distinct sharers of the world—western capitalism, fundamentalist Islam, and Asian neo communist or dictatorial societies.

I do not think so. Rather I would suggest a much quieter but far more extensive revolution. It is the revolution of the cell phone and micro moveable technology. By 2009 there were more than 3000 million cell phones in use. It is just under one cell phone for every two people in the world. And if you subtract the number of children in the world, a dangerous assumption, there is almost one per person. (actually somewhat over three per four people).

The pc and the internet changed the world in the twentieth century. However, even now there are already three times more mobile phones than pcs. Almost 90% of all mobile phones are sold with web browsers and the word "app" is known from the southern tip of Tierra del Fuego to the Northern tip of Kamchatka.

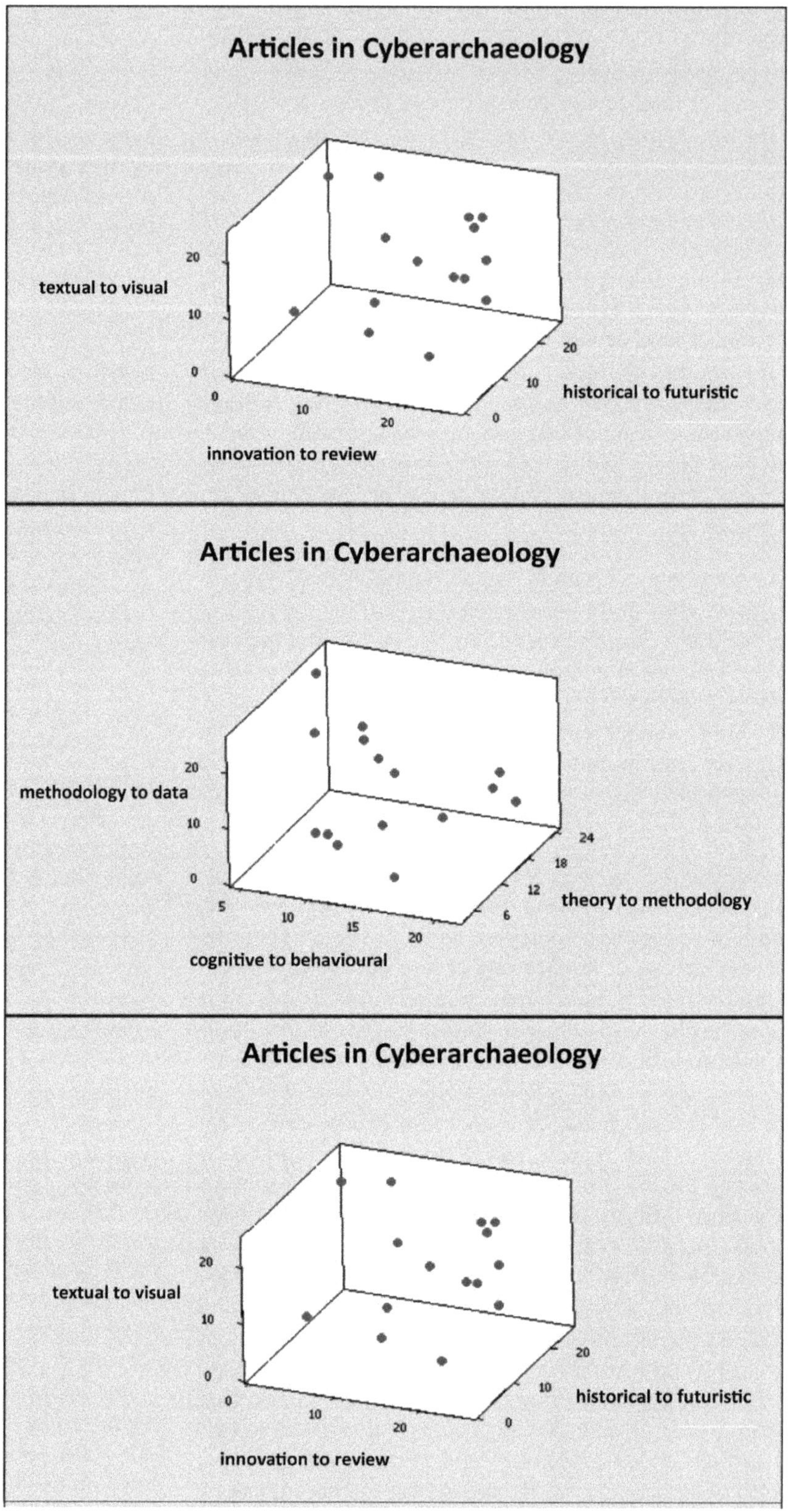

Figure 1. A variety of axes on which to arrange articles

This book clearly shows the progression of archaeological thinking in the late twentieth and early twenty first century. In it if one looks carefully, one can see the beginnings of the twenty first century. As Cyberarcheology was relevant for a cyber age, I-archaeology will be relevant for an i-world. It will be:

- Mobile
- completely interconnected from every individual to every individual at all places all times
- visual and sound
- versatile with numerous standardized and individualized apps
- immediate

It will change the way people conceive of the past for they will be connected to all of it all of the time if they so wish.

Article	Problem and Methodology	Success
Maase, Forte, Janecky, Barrientos	Create virtual reality tools with interactive variables such as climate and topography to examine meteor catastrophes as evolutionary events	Success is educate the public.
Kurillo, Forte, Bajcsy	Prototype collaborative real-time interaction with 3D archeological models using video streaming and teleimmersion to create avatars. The avatars interact virtually with remote archaeological site. Tools include avatar navigation, measurement, pointing, annotation, object moving, mode manipulation.	Success is adequacy of the avatar interaction to reflect reality and or to provide a test bed of new ideas. As a prototype success is to facilitate the study and analysis of a virtual reconstruction process in archaeology using virtual community to re-contextualize and reassemble spatial archaeological data.
Kirsch	Cognition is dependent upon tangibly interacting with material objects thus virtual realities must go beyond the visual to the tangible. Present avatar relationships to virtual data are not equivalent to the real. The hefting of a virtual object is not the same as hefting a real object.	Success is determine criteria for equivalency and demonstrate them
Kenderdine	Uses immersive installations Hampi Place and Hampi Alive to demonstrate how post processual archaeology through performance theory and narrative create a trichotomy between performers, spectators, and place.	Success is the experience of Hampi
Galcazzi, Di Giuseppantonio Di Franco, Dell'Unto	Use 3D virtual cyber mind maps to interpret the religious, spatial, and symbolic relationships through afforandances in Western Han Mural Tombs. Using a five step process they add 3dimensionl reconstruction, to texturized drawings made by archaeologists, to overlapped drawings of the paintings, adding cybermaps, and repositioning original picture.	Success is new way to rethink and rearrange the old way to make archaeological documentation with hybrid learning among symbolic ,reconstructive and motor processes. The user reads and decodes textual symbols and recreates a mental visual perception. Secondarily, the actions can be seen to create an effect.
Champion	Need to create term and methodologies to bypass individual artifacts and conserve the situated knowledge of retrieval. Navigating through virtual environments with appropriate cognitive maps and tools will create adequate cultural immersion. Many examples provided.	Success is adequate cultural immersion so that navigation does not interfere with the cultural learning goals of cyber archaeology
Levy, Petrovic, Wypych, Gidding, Knabb, Hernandez, Smith, Schuluz, Savage, Kuester, Ben-Yosef, Buitenhuys, Barrett, Najjar, DeFanti	On site digital archaeology (OSDA) is an enhanced system with a core GIS with an active cyberinfrastructure for Levantine archaeology. Digitally enabled archaeology is obtained through such techniques as total stations, GPS, Lidar, airborne imagery, and 3 D artifact scanning. How one should synthesize this diverse data is shown.	Success is an integrated digital set of diverse data that is ready for a variety of analytical techniques.
Clark	Examining the role of cyber archaeology in processual and post processual archaeology, he determines that archaeology is construction of models and sites and not the fallacy of reconstruction.	Success is determined by having criteria for models that allows users to determine the levels of accuracy.

Barcelo	Examines the theory and design of an "robotic" or "automatic archaeologist"	Success is emulating behavior productively not constructing a "thinking" robot
Van der Elst	Cognitive landscapes should not assume the cross cultural universality of time and space concepts.	Success is adequate time and space translators.
Lercari	Strives for a new methodologies of cultural transmission showing that 3D graphics, real time engines, and interactive story telling are fundamental resources for representing ancient urban landscapes.	Success is cultural transmission to the public

Table 1. Problems, Methodologies and Criteria of Success for Cyberarchaeology articles

References

Allen, K. M. S., S. W. Green, et al. (1990). *Interpreting space: GIS and archaeology.* London; New York, Taylor & Francis.

Ashby, W. R. (1956). *An introduction to cybernetics.* New York, J. Wiley.

Binford, L. R. (1977). *For theory building in archaeology: essays on faunal remains, aquatic resources, spatial analysis, and systemic modeling.* New York, Academic Press.

Binford, L. R. and G. I. Quimby (1972). *An archaeological perspective.* New York, Seminar Press.

Clark, G. (1961). *World prehistory, an outline.* Cambridge [Eng.], University Press.

Clark, G. (1969). *World prehistory: a new outline.* London, Cambridge U.P.

Clarke, D. L. (1968). *Analytical archaeology.* London, Methuen.

Costopoulos, A. (2001) Evaluating the Impact of Increasing Memory on Agent Behaviour: Adaptive Patterns in an Agent Based Simulation of Subsistence. *Journal of Artificial Societies and Social Simulation* 4.

Hodder, I. (1986). *Reading the past: current approaches to interpretation in archaeology.* Cambridge [Cambridgeshire]; New York, Cambridge University Press.

Hodder, I. (1989). *The Meanings of things: material culture and symbolic expression.* London; Boston, Unwin Hyman.

Hodder, I. (2001). *Archaeological theory today.* Cambridge, UK, Polity Press.

MacKay, D. (1954). "On Comparing the Brain with Machines;" *The Advancement of Science* 40 (March 1954): 402-406.

Massachusetts Institute of Technology. Computation Center, J. McCarthy, et al. (1960). *LISP I programmer's manual. Cambridge,* Massachusetts Institute of Technology, Computation Center and Research Laboratory of Electronics.

McCarthy, J., Massachusetts Institute of Technology. Artificial Intelligence Laboratory, et al. (1959). *Recursive functions of symbolic expressions and their computation by machine: the LISP programming system.* [Cambridge, Mass], Artificial Intelligence Group, M.I.T. Computation Center and Research Laboratory of Electronics.

McCulloch, W. and W. Pitts (1943). "A logical calculus of the ideas immanent in nervous activity." *Bulletin of Mathematical Biophysics,* 5: 115-133

Minsky, M. L. (1963). *Neural nets and theories of memory.* [Cambridge, Mass.], Massachusetts Institute of Technology, Project MAC.

O'Keefe, P. J. and Institute of Art and Law (Great Britain) (2007). *Commentary on the UNESCO 1970 Convention on the means of prohibiting and preventing the illicit import, export and transfer of ownership of cultural property.* Powys, England, Institute of Art and Law.

Rosenblatt, F. (1962). *Principles of neurodynamics; perceptrons and the theory of brain mechanisms.* Washington, Spartan Books.

Samuel, A. L. (1995). *Some Studies in Machine Learning Using the Game of Checkers. Computation and Intelligence: Collected Readings.* G. F. Luger. Menlo Park, CA/Cambridge, MA/London: AAAI Press/The MIT Press.

Selfridge, O. G., E. L. Rissland, et al. (1984). *Adaptive control of ill-defined systems.* New York, Published in cooperation with NATO Scientific Affairs Division [by] Plenum Press.

Shanks, M. and C. Y. Tilley (1987). *Social theory and archaeology.* Cambridge, Polity in association with Blackwell.

Shanks, M. and C. Y. Tilley (1992). *Re-constructing archaeology: theory and practice.* London; New York, Routledge.

Standardization, I. O. f. (2006). *Information and Documentation-A reference ontology for the interchange of cultural heritage information.* Geneva Switzerland, International Organization for Standardization ISO 21127:2006 108.

Tilley, C. Y. (1993). *Interpretative archaeology.* Providence, Berg.

Tilley, C. Y. (1999). *Metaphor and material culture.* Oxford, UK; Malden, Mass., Blackwell Publishers.

Turing, A. M. and B. J. Copeland (2004). *The essential Turing: seminal writings in computing, logic, philosophy, artificial intelligence, and artificial life plus the secrets of Enigma*. Oxford; New York, Clarendon Press.

Viterbi, A. J. (1995). *CDMA: principles of spread spectrum communication.* Reading, Mass., Addison-Wesley Pub. Co.

Viterbi, A. J. and J. K. Omura (1979). *Principles of digital communication and coding.* New York, McGraw-Hill.

Wiener, N. (1948). *Cybernetics; or, Control and communication in the animal and the machine*. Paris, Cambridge, Hermann et Cie; Technology Press.

Winograd, T. (1983). *Language as a cognitive process*. Reading, Mass., Addison Wesley Pub. Co.

Winograd, T. and F. Flores (1986). *Understanding computers and cognition: a new foundation for design*. Norwood, N.J., Ablex Pub. Corp.

Winograd, T. A. (1972). *Understanding natural language*. New York, Academic Press.

Introduction to Cyber-Archaeology

Maurizio Forte
University of California, Merced
mforte@ucmerced.edu

This book collects articles from two different workshops organized in 2009 and 2010. The TAG (Theoretical Archaeology Group) Conference in Stanford (May, 1-3, 2009) - workshop on Cyber-archaeology - and the conference *Diversifying Participation. Digital Media and Learning* - workshop on Virtual Collaborative Environments for Cultural Heritage (February, 18-20, 2010 in San Diego, La Jolla).

The workshop organized for TAG was aimed to analyze the epistemology of cyber-archaeology in relation with the research topics of the conference: state of the art, methods, theory, applications and overviews. The workshop in San Diego was focused mostly on collaborative environments, collaborative research, virtual models and simulation studies. Both initiatives aimed to analyze the state of the art in the field of virtual-cyber archaeology, collaborative research and immersive environments. The participation in these meetings was multidisciplinary: archaeologists, computer scientists, historians, cognitive scientists and art historians have presented and discussed very interesting contributions from different perspectives.

Thanks to the collaboration of Professor Thomas Levy, at UC San Diego, we had the chance to extend our discussions even in the immersive cave (StarCave, CALIT 2), in front of a 3D simulation of an archaeological excavation (the Iron Age site of Khirbat en-Nahas, ca. 1200 – 900 BCE, copper production center in Jordan). The StarCAVE is a 5-walled, rear-projected, 360 degrees virtual reality device. The discussion into the virtual cave has further demonstrated the complexity of a cyber-archaeological approach and also the difficulties to describe this kind of virtual phenomenology.

The general scope of this book is nevertheless to introduce and discuss current phenomenology of *cyber-archaeologies* and related applications at epistemological, technological and methodological level through some significant theoretical approaches and case studies. Despite this attempt is incomplete and very preliminary, the challenge is to design a cultural manifesto founding this work from different perspectives and with different multidisciplinary contributions, and theoretical debates. It is still very difficult to introduce and define possible borders, weaknesses and potentialities of a new research field specifically in relation with new technologies, highest levels of embodiment, immersive systems and 3D social networks. Is this idea to design such new field of research really sustainable? How can we say that this post-modern revolution is more cyber than virtual? What happens in this digital era?

The rapid and incremental use of 3D digital technologies in archaeology in the field and labs is able to design diverse and uninvestigated workflows in the generation, representation and communication of data. This digital migration of data and models in so diverse domains creates unexpected results and more advanced knowledge. According to Bateson's thought (Bateson, 1972) we can define this process the *map-code* of the cybernetic cycle. The study of the code is mandatory for re-analyzing the interpretation process in the light of a cybernetic perspective: the feedback created by different interactors operating in the same environment/ecosystem generates further feedback and not pre-determined inter-connections.

I want to insist on the importance of non-predetermined conditions in the evaluation of a cyber-process, because this should be the core of our investigation: how do we generate knowledge by cyber simulations in a digital ecosystem? How do we formulate an interpretation when all our models of knowledge coexist in a digital-cyber domain?

The core of the process is not into the model, data or environment but in the mutual relations produced by interaction/embodiment/enaction (Maturana and Varela). The enactive cognitivism discusses the dichotomy between intern and extern: cognition is an action "embodied" (Varela *et alii* 1991). In terms of enaction, the cognition depends on perceptual-motor experience and these capacities belong to biological-genetic but also cultural contexts. Thus the acquisition of data from the environment would be identified in the circularity between action and experience and between action and knowledge (Varela *et alii* 1991). Every perceivable information in the environment comes from a perceptual-motor interaction. The object takes shape because of our eco-activity and therefore we and the object take shape together (Varela, 1999, 66). In conclusion, even the cyber interaction or behavior in a virtual environment can be totally reconsidered as an exchange of information organism-environment.

The basic principle is that that perception and knowledge of the environment are ruled by affordances (Gibson, 1979) and these are developed during the simulation process. The affordance represents the virtual link between object and environment: the code for re-interpreting the past in space and time. The multiplicity of the affordance is addressed to perceive and think the object as *enactive knowledge*; it is a dynamic knowledge because it is defined by simulation. For example a figurine or a jar can have ritual meaning or can be just an ornament or a simple object used in the daily life; the interpretation code is given by the affordance but all the affordances can change through time and space and by the environment. Again an object can have different uses in the same space or context or in different spaces and contexts in a diachronic evolution. The connection with and between objects and environment can design the affordance as potential "event". In the Gibson's idea (Gibson, 1979) the affordance comes from the perception: it semantic value identifies the human action or thought. The context is the code for understanding the affordance, therefore there is not a right or wrong representation "a priori" but the interpretation comes from the enactive process. The original shape, use or project around an object could be totally different from its later standardized use.

- If we consider these premises, it is quite difficult to accept the term" reconstruction of the past", because it recalls a peremptory idea to generate a "pre-defined" knowledge and not, on the contrary, a "possible" knowledge. A possible knowledge comes from a simulation of the past and not from its reconstruction.
- This point is very important in the discussion about cyber-archaeology because it focuses on the potentiality-virtuality of the interpretation vs. the actuality of the physical world. In the cyberarchaeological process the focus is the simulation, that is the enactive-dynamic behavior of the virtual actor and the digital ecosystem. In consequence of this, the workflow able to move and migrate data from the fieldwork to a simulation environment can generate different affordances and cybernetic models: each one can create a feedback and this is a new map-code for the interpretation. In this way we can summarize the following key points:
- Cyber Archaeology can represent today a research path of simulation and communication, whose ecological-cybernetic relations organism-environment and informative-communicative feedback constitute the core. Archaeology is entirely part of the contemporary society and it is the gateway to the ancient world; it cannot represent the ancient world. The cyber process creates affordances and through them we are able to generate virtual worlds by interactions and inter-connections.
- Virtual Archaeology was mainly visual, static, graphic and oriented to photorealism
- Cyber-archaeology is not necessarily visual, but dynamic, interactive, complex, autopoietic and not necessarily oriented to photorealism.
- The past cannot be reconstructed but simulated. The cybernetic simulation develops affordances and there relations constitute the potentiality of the interpretation process.
- Potential past. It is more appropriate to think about a potential past, a co-evolving subject in the human evolution generated by cyber-interaction between worlds.
- The validation of this process depends on the relationships between present and past societies.
- Cyberarchaeology is the process of reconstruction-simulation. The archaeology of the third millennium is able to process, interpret and communicate much more data and information than in the last two centuries.
- In the 90's most part of research projects in virtual archaeology were visual oriented; now we think that in the third millennium they will be cybernetic-oriented.
- The capacity to transmit knowledge and interpretation depends on a complexity of diverse factors: technology, format, accuracy, induction-deduction, communication, context, ontologies, etc. Are we aware of how much data can be produced and disseminated in this era? And how much fast is this process?
- The virtual ontology of archaeological information, or the cybernetics of archaeology, refers to all the interconnective relationships which the datum produces, the code of transmission, and its transmittability.
- The datum is never neutral and consequently, we have to improve the properties of the affordances.

In short, we can imagine the following cybernetic syllogism: if the knowledge depends on the affordances, even our perception does; more affordances multiply the level of interpretation and generate more feedback; more feedback and increased perception makes more difference in cybernetic sense (Bateson, 1972) and then produce/exchange more information between observers and environment.

For describing much more in detail this process I want to use the theory of *autopoiesis* of Maturana and Varela, and the metaphors of rhizome and puppets by Deleuze and Guattari.

"*Autopoiesis* is the process whereby an organization produces itself. An autopoietic organization is an autonomous and self-maintaining unity which contains component-producing processes (Principia Cybernetica Web)". The autopoietic process can describe properly what can happen in a cyber-archaeological process: the creation of information and knowledge is not pre-determined, it is not an unilinear process. On the contrary, it is the effect of multiple interactions, connections, feedback; it is a presentation of a multiplicity. In the autopoiesis the cybernetic system can generate knowledge independently from the actors involved; if we study the code, we can find the interpretation.

An autopoietic machine is a machine organized (defined as a unity) as a network of processes of production (transformation and destruction) of components which: (i) through their interactions and transformations continuously regenerate and realize the network of processes (relations) that produced them; and (ii) constitute it (the machine) as a concrete unity in space in which they (the components) exist by specifying the topological domain of its realization as such a network. (Maturana, Varela, 1980, p. 78).

Cyber-archaeology operates as an autopoietic machine, able to manage networks of processes in non-predetermined spaces of knowledge. If for Maturana and Varela, the unit was the biological cell, in the cyber process a unit can be a digital cell, an organism or its representation. Every digital cell can have multiple non-predetermined affordances.

Autopoietic avatars, virtual humans, artificial lifes can populate the cyberspace, distributing complex actions and behaviors in different realms: the reality (for example the archaeological excavation) and the cyberspace. Reality and cyberspace will become hybrid units: a new project with UC Merced and UC Berkeley (supported by CITRIS) is working on the creation of cyber-archaeological collaborative spaces: virtual humans, labs and archaeologists could collaborate on site, generating 3D data in real time.

At this point it is interesting to recall the metaphor of puppets and strings: *puppet strings, as a rhizome or multiplicity, are tied not to the supposed will of an artist or puppeteer but to a multiplicity of nerve fibers, which form another puppet in other dimensions connected to the first: "Call the strings or rods that move the puppet the weave. It might be objected that its multiplicity resides in the person of the actor, who projects it into the text. Granted; but the actor's nerve fibers in turn form a weave. And they fall through the gray matter, the grid, into the undifferentiated...* (Deleuze & Guattari, 1987 in *A Thousand Plateaus: Capitalism and Schizophrenia* [University of Minnesota Press, 1987].

In this metaphor we have independent and dependent factors in the dynamics of puppets. The dependent factors are the hands moved by human actors, the independent factors are the multiplicity-whole of the strings that, by mutual interactions, generate unpredictable movements and dynamics. In these terms they are a rhizome and a rhizome is also a map. *"The rhizome is altogether different, a map and not a tracing. What distinguishes the map from the tracing is that it is entirely oriented toward an experimentation in contact with the real. The map does not reproduce an unconscious closed contact with the real. The map does not reproduce an unconscious closed in upon itself; it constructs the unconscious. It fosters connections between fields, the removal of blockages on bodies without organs, the maximum opening of bodies without organs onto a plane of consistency. It is itself a part of the rhizome. The map is open and connectable in all its dimensions; it is detachable, reversible, susceptible to constant modification.* Deleuze & Guattari in A Thousand Plateaus: Capitalism and Schizophrenia [University of Minnesota Press, 1987].

The interpretation of the past as a rhizome and a map, re-launches the discussion about cyberarchaeology, with additional key points:

3D information can be regarded as the core of the knowledge process, because it creates feedback, then cybernetic difference, among the interactor, the scientist and the ecosystem.

It is argued that Virtual Reality (both offline and online) represents a possible ecosystem, which is able to host top-down and bottom-up processes of knowledge and communication.

In these terms, the past is generated and coded by "a simulation process".

What is the relationship between information and representation? How much information does a digital model contain?

In the research project on the Western Han tombs, whose preliminary application is here presented (Galeazzi, Di Giuseppantonio, Di Franco Del'Unto), for example, we are experimenting different non-symbolic codes associated to the mural paintings and the iconography of the monuments. The 3D cyber-map represents an attempt to re-elaborate the interpretation of the frescos using non-Western codes and stressing the importance of 3D visual and narrative connections for a multivocal archaeological interpretation. The idea to represent the tomb as a macrocosm is propaedeutic to understand its cosmology in the Western Han period by a visual language. The cyber-map designs the network of interpretation codes and cultural contexts, otherwise not understandable in a linear sequence (spatial, temporal or narrative).

If we see this cyber-era in a retrospective sense, we have to compare the evolution of digital archaeology from the processualism to the post-processual thinking and, as Zubrow remarks in his introduction, with the scope to analyze also hybrid forms of both approaches.

From the digital point of view the first outcome of the archaeological processualism was the use of statistical processing and quantitative methods in different domains, mathematics, geography, archaeometry, anthropology, archaeology and related disciplines. The critique of subjective methodologies pointed out the need of hyper-taxonomies for interpreting the past and this computing archaeology seemed a tangible and sustainable way for the processualist dream: an objective "scientific" interpretation. The idea that computing was able to make data and information more "objective" stimulated the archaeological community to develop specific methodologies in labs and post-processing, but not so much in the field. The processual dream was to define borders and objective aspects of the discipline also in relation with the new capacity of computing. I think that the idea and wish to accelerate the informational process (by quantitative methods) were strongly correlated with the power of computing (firstly by mainframe, then by personal computers). Unfortunately too much attention was given to the capacity of the process and less to the cybernetic content of the models. In short, the simulation was concentrated in the use of digital algorithms and not enough on the study and representation of the models. We could say "more numbers than knowledge": the indiscriminate use of taxonomies without critical discussions, can cancel any chance to recontextualize the interpretation, to imagine the multiplicity and rhizome of the knowledge.

A similar situation of methodological hyper-descriptive re-discovery is identifiable with the birth of virtual archaeology: virtual worlds able to visualize the past with a very impressive photorealism which I called "the wow factor" in a recent paper (Forte, 2009...). This digital 3D seemed to suggest extraordinary revolutions in the archaeological field, more specifically in relation with the communicational aspects. In the same time virtual reality was recognized as part of the future human social life with incredible expectations from the scientific and media world. The big bang of virtual archaeology in the 90s, determined by the development of computer graphics techniques, was in part very superficial: very few applications were accessible and validated by a cybernetic workflow; just a minimal percentage of data migrated to collaborative processing or shared environments. The expectations of a radical revolution were disappointed and also the idea of a systematic and standardized use of virtual reality in archaeology seemed to trace a very difficult path.

It is interesting to note that the first virtual worlds in archaeology were born at the beginning of the 90s not in academic labs, universities or research centers, but by the initiative of corporations, industries and multinational companies (for example, Taisei Corporation in Japan, Forte 1996, 2000...). This happened, substantially, because of the economic and technological power of these stakeholders, by the dream of some industries to be pioneers in the fields, but followed by a significant skepticism of the academic world. Spectacular and esthetic factors obscured a serious discussion about the process: how is it possible to reconstruct the past? How can we evaluate and test models and methods? What happens in this black box? Similar questions were circulating just in a few rooms of the academic world, with a limited impact in broader general discussions. At the end of the 90s, after the missing revolution of virtual archaeology (at least in comparison with the expectations), the strong impact of 3D technologies (3D scanners, photogrammetry, computer vision, etc.) in the fieldwork introduced the third era of digital archaeology (the first quantitative, the second virtual), we could call cyber. The production of impressive amount of data in a very short time placed additional research questions but not in the right directions. The focus was much more oriented in the applications and in the power of digital technologies instead of serious and deep analysis on the computational models and 3D environments.

After almost 15 years of debates on the role of computing and virtual applications in archaeology, do we need to rethink the definition of the controversial, popularized and now accepted domain of "virtual archaeology" (Forte, 1996)? Is this sufficient for defining what happens in the complex scenarios of the virtual worlds? Is this process validated by a specific workflow?

At the time of the first book "Virtual Archaeology", it was very controversial to imagine a field of study, research and education under this name and, in any case, the discussion was focused on the role of VA as visual (mostly) application for the reconstruction of the past. These categories of visual re-launched the actuality of the reconstruction of the past, as something digitally "tangible", confutable and transmittable (Forte, 2008, 2009). However, in the last decade virtual archaeology was academically recognized as a research field, with specific conferences, workshops, exhibits and even university courses. The debate of the role of virtual archaeology was mainly oriented on the representational aspects of the models and very few on the autopoietic effect of interactions, design and interfaces. Virtual archaeology was not able, substantially, to elaborate sophisticated cyber behaviors in 3D real time environments.

Actually there was a relevant discrepancy between bottom-up and top-down processes. The phase of data collecting, data-entry (bottom-up) was mostly 2D and analogue, while the data interpretation/reconstruction (top-down) was 3D and digital. How was it possible to conjugate both methods? How was the scientific workflow? The phase of data collections-data recording should be totally integrated in the simulation-reconstruction process; if we separate the two domains (bottom-up/recording, top-down/reconstruction/interpretation) we lose information and the capacity to compare and to validate the data workflow in the virtual environment.

In the first "analogue" phase (80s-90s) the 3D models were based principally on manual drawings, meta-data or made by cultural comparisons, imagination, pure graphical hypotheses. There were a few markers of the spatial data and archaeological remains in the 3D reconstruction which was a sort of evocative virtual world (and not on scale). In particular, the acceptance of an archaeology "virtual" and not "just digital" was depending on the capacities and potentialities expressed by technologies, but, first of all, by the introduction of new methods and approaches, starting from the fieldwork and arriving to the final 3D communication process. The invention of VRML (Virtual Reality Modeling Language) as 3D Web language, the migration of 3D rendering and modeling from mainframes to personal computers, were showing new and more advanced capacities for the reconstruction/visualization of the past. In the 90s this reconstructive approach was imaging mainly a single utopian past and not a multivocal open interpretation. Was that a real revolution? Was it able to change radically the methodology and the interpretative process in the dialectic relation between fieldwork-data capturing and communication/representation? In other terms we could say that this was a wish, and if Virtual Archaeology was a missed revolution, however, it was able to push the archaeologists to ask new research questions.

In the progression of studies and research in digital-computing applications in archaeology I find some analogies on what happened in the 60's with the New Archaeology and with the "boom" of the statistical methodologies/analyses in the 70s and 80s. The idea is to find some communality not in the historical-methodological context and background, but in the process of trusting the methodologies-technologies available, without a serious epistemological debate.

Therefore it is appropriate to compare virtual (VA) and cyber archaeology (CA) trying to understand the differences of these two important periods of scientific research in archaeology. VA was mostly visual (partially analogue), graphic and representational; CA is totally digital, immersive, autopoietic, interconnective, based on affordances. If in VA the keyword is "seeing", in CA is "playing". This difference is very important, since now we are able to transmit and distribute much more digital-cyber knowledge than in the past.

Information is not innocent and neutral, whatever we produce, is based on the way we perceive it. This knowledge can be seen as a process of selection: "selection filters the noise and reveals at the same time the information, channeling it into patterns that will result in understanding" (Taylor 2005: 261). To select information we must process it in accordance with a valid sequence of acquisition, modeling, representation, and communication. In this sense it is possible to distinguish between accumulated (i.e., not selected) knowledge and the quality of perceived (selected) information (Forte, 2009). The increased information, its "density" constitutes the focus of a research path of virtual archaeology.

Cyber Archaeology can represent today a research path of simulation and communication, whose ecological-cybernetic relations organism-environment and the informative-communicative feedback constitute the core, but they have to be still fully investigated (Forte, 2007).

References

Binford L.R. - Binford S. R., *New Perspectives in Archaeology*, Chicago 1968

Bateson, G., 1972. *Steps to an Ecology of Mind* (Chandler Press: San Francisco).

Bateson, G., 1979. *Mind and Nature. A Necessary Unit* (Dutton: New York).

Fletcher M. - Lock G.R., *Digging Numbers*, Oxford 1991

Forte, M., 1997 (ed.). *Virtual Archaeology* (Harry Abrams Inc. Publishers: New York).

Forte, M., 2000. "About Virtual Archaeology: Disorders, Cognitive Interactions and Virtuality," in Barcelo J., Forte M., Sanders D., 2000 (eds.), *Virtual Reality in Archaeology, BAR International Series* S843 (Oxford, BAR Publishing) 247-263.

Forte, M., 2003. "Mindscape: Ecological Thinking, Cyber-anthropology, and Virtual Archaeological Landscapes," in *The Reconstruction of Archaeological Landscapes through Digital Technologies,* edited by M. Forte, P. R. Williams, *Proceedings of the 1st Italy-United States Workshop, Boston, Massachussets, USA, November 1-3, 2001, BAR International Series* 1151, (BAR Publishing: Oxford) 95-108.

Forte M., Pietroni E., 2009, *3D Collaborative Envirionments in Archaeology: Experiencing the Reconstruction of the Past*, in "International Journal of Architectural Computing", Issue 01, Vol. 07, March 2009, 57-75.

Gibson, J, 1950, The perception of the Visual World, Greenwood Press, Westport CT, 1950.

Harrison, S., Dourish, P., 1996. "Re-place-ing Space: The Roles of Place and Space in Collaborative Systems," *Proceedings of the ACM 1996 Conference on Computer Supported Cooperative Work, Boston, Massachusetts, United States*.

Levy, P., 1995. *Qu'est-ce que le virtuel?* (Editions la Decouverte: Paris).

Maturana H., Varela F., 1980. "Autopoiesis and Cognition: The Realization of the Living," *Boston Studies in the Philosophy of Science*, edited by Robert S. Cohen, Robert S. and Max W. Wartofsky, vol. 42.

Moscati P., *Archeologia Quantitativa: nascita, sviluppo e "crisi"*, "Archeologia e Calcolatori", 7, 1996, pp. 579-590

Taylor, M.C. 2005. *Il momento della complessità. L'emergere di una cultura in rete* (Codice edizioni: Turin).

From Computable Archaeology to Computational Intelligence New Prospects for Archaeological Reasoning

Juan A. Barceló
Universitat Autònoma de Barcelona. Spain

Abstract

The question of whether it is possible to automate the archaeological knowledge production is of both great theoretical interest and increasing practical importance because knowledge and information are being generated much faster than they can be effectively analyzed. The approach adopted here is based on a fact that archaeologist couldn't evaluate 15 years ago: Computer programs do work in real science, not only in archaeology. Maybe they are more successful in other "harder" sciences, but we cannot deduce from this fact that Archaeology *is* a *different* kind of science. Computable archaeology –if you do not like the expression "automatic archaeology" is the proper way of exploring new ways of answering the questions we have not yet answered.

Key words: *Reasoning, Artificial Intelligence, Automatic Explanation, Expert System, Knowledge base, Machine Learning*

1 "Automatic" Archaeology: a useless endeavor, an impossible dream or reality?

Is it possible to build a machine to do archaeology? Will this machine be capable of acting like a scientist? Will this machine be capable of understanding how humans act, or how humans think they acted in the Past? These questions are rather original in Archaeology. They are even provocative, given the current fashion of post-modern and hermeneutic approaches. Nevertheless, the dream of an automated archaeology has existed latently in some hidden places of the archaeological world.

We should imagine an *automated* or *artificial archaeologist* as a machine able to act as any of us, human archaeologists, learning through experience to associate archaeological observations to explanations, and using those associations to solve archaeological problems. It should have its own "cognitive core" and should interact with the world to make changes or to sense what is happening. It should have self-awareness, and it should run autonomously. This last requirement indicates that such a machine should operate, self contained, under all reasonable conditions without recourse to a human operator. Autonomy means that a robot can adapt to changes in its environment or in its very nature and continue to reach its goal (Murphy 2002, pp. 3-4).

In a sense, our automated archaeologist may be seen as a *cognitive robot*. Ronald Arkin gives the following definition: "an intelligent robot is a machine able to extract information from its environment and use knowledge about its world to move safely in a meaningful and purposive manner" (Arkin 1998, p. 2, see also Stein 1995). By substituting "moving" by "solving archaeological problems", we get a perfect working definition for our automated archaeologist. "Intelligent" implies that the robot does not work in a mindless, repetitive way; it is the opposite of the connotation from factory automation. An artificial archaeologist should be then a *"physically instantiated... system that can perceive, understand ... and interact with its environment, and evolve in order to achieve human-like performance in activities requiring context (situation and task) specific knowledge"* (European Commission Framework 6 objective for 'Cognitive Systems').

It is important to realize that any formal definition of an *automaton* includes both human beings and the so called "intelligent" robots within the same reference class. In the specific case of scientific method, thinking can be construed as manipulating or processing information, whereas performing computations on strings of digits is the way the computer process information. *Automata* are devices with defined inputs, outputs and structure of inner states, which perceive the *environment* through some mechanisms called *sensors* and act upon that environment through other mechanisms called *actuators* or *effectors* (Aleksander and Morton 1993, p. 97). Here I am assuming that *explanations* are for our automatic archaeology machine a form of *behavior (*Donahoe and Palmer 1994, Kirsch and Maglio 1995). If the human archaeologist has eyes, ears, and other organs for sensors and a brain, together with other body parts for actuators, then robotic archaeologists might have cameras and infrared range finders for sensors and a computer program for cognitive activity. A human has input senses, a muscular output and a cognitive ability to use this to great effect (excavating old sites, drawing finds, interpreting data, teaching archaeology, etc.). The automated archaeologist receives numbers, file contents as sensory inputs and acts on the environment by displaying on the screen, writing files. Although a human has an inner structure of wondrous complexity, which is the seat of consciousness and intelligence, mathematically even the best of us can be described as an automaton.

May be some of you will say that we do "not yet" have automatic archaeologists, but we should hurry up to the engineering department and build them for having someone able to substitute us in the tedious task of studying ourselves and our past. Other readers will claim: "fortunately, such a machine will never exist!" "Why we need such an awful junk? Computers cannot *emulate* humans". These critics seem to think that computer programs are guilty of excessive simplification, of forcing knowledge, or distorting it, and of failing to exploit fully the knowledge of the expert, but it seems to me that it is Archaeology, and not computer programs, what is "narrow minded". The saddest thing is that archaeologists do not know how they know archaeological matters.

The so called "intelligent" machines incite instinctive fear and anger by resembling ancestral threats -a rival for our social position as more or less respected specialists. But robots are here, around us. I've never heard of a claim against wash machines selecting "intelligently" the best way to wash a specific tissue, or a photo camera with an "intelligent" device measuring luminance and deciding by itself the parameters to take the picture. So, why to have fear of a machine classifying a prehistoric tool and deciding "intelligently" its origin, function and/or chronology? Rather than arguing whether a particular behavior should be called intelligent or not -a point that is always debatable- I try to provide answers to the following question: Given some archaeological explanation that we find interesting in some ways, how does the explanation come about? Rather than using intuition as the sole guide for formulating explanations of past human behavior, we need a theory of why a specific computation or a group of related computations should be performed by a system that has certain abilities.

The discussion is between what is considered an *artificial* way of reasoning (computer programs) and a *natural* way of reasoning (verbal narrative). Critics of computationalism insist that we should not confound scientific statements with predicate logic operations, since discursive practices or argumentations observed in a scientific text are not "formal". By that reason, they are tributary, to a certain extent, from the Natural Language and the narrative structure (literary) of which scientific texts derive. I take the opposite approach: scientific problem solving stems from the acquisition of knowledge from a specific environment, the manipulation of such knowledge, and the intervention in the real world with the manipulated knowledge. The more exhaustive and better structured the knowledge base, the more it emulates a Scientific Theory and the easier will be the solution to the scientific problem, and more adequate the interpretations we get.

2 Machines who think

The dream of an "intelligent" machine is very old in the history of Philosophy, and it relates with the progressive discovery that nature and human acts are not divine secrets, but could be broken down and understood systematically. Since Socrates, philosophers have often anthropomorphized a problem by imagining a *demon* accomplishing a task that was difficult to understand but seemed to be possible. In some cases, such a demon had a kind of mechanical nature. Maybe the most relevant "precedent" of computational intelligence was the logic machine for discovering nonmathematical truths through combinatorics, called the *Ars Magna*, imagined by Ramon Lull, a medieval mystic (13th. century), and one of Catalonia's greatest poets. It was more than three centuries after the *Ars Magna* that it influenced Thomas Hobbes (1588-1679), who stated "by ratiocination, I mean computation. Now to compute is either to collect the sum of many things that are added, or to know what remains when one thing is taken out of another. Ratiocination, therefore, is the same with addition and subtraction". Hobbes wanted to account for all cognitive activities of the mind in terms of computation, and computation is cashed out in terms of manipulation (transformation) of computable entities

Gottfried Wilhelm von Leibniz (1765) envisioned a universal calculus of reasoning by which arguments could be decided mechanically. "Everything proceeds mathematically...if someone could have a sufficient insight into the inner parts of things, and in addition had remembrance and intelligence enough to consider all the circumstances and take them into account, he would be a prophet and see the future in the present as in a mirror". He seemed to see the possibility of mechanical reasoning devices using rules of logic to settle disputes.

Many of the Enlightenment natural philosophers took similar views. Julien Offray de La Mettrie (1709-1751) was may be the first who suggested that "man is a machine", more as a metaphor than as a mechanical possibility. We should wait until 20th century for a new turn of the screw. In 1914, Bertrand Russell imagined recording instruments that could perceive the world in place of a human observer. Russell's virtual observer eliminates the subjectivity of perception of what things really are. "There is no theoretical limit to what can be done to make mechanical records analogous to what a person would perceive if he were similarly situated" (Russell 1959).

With early twentieth century inventions in electronics and the post–World War II rise of modern computers, possibilities gave over to demonstrations. As a result of their awesome calculating power, computers in the 1940s were frequently referred to as "giant brains." As a consequence, in the middle of the 20th century, a handful of scientists began to explore a new approach to thinking machines based on their discoveries in neurology, a new mathematical theory of information, the engineering approach to control and stability mechanisms, and by the availability of machines based on the abstract essence of mathematical reasoning. Turing's 1950 seminal paper in the philosophy journal *Mind* was a major turning point in this history. The paper crystallized ideas about the possibility of programming an electronic computer to behave intelligently, including a description of the landmark imitation game that we know as Turing's Test.

Cybernetics preceded Computational Intelligence as a discipline that shared the ambition of providing an explanation of the nature of higher cognitive activities such as learning, memory, and more generally, teleological purposive behavior, couched in the terminology of the physical sciences. The term *cybernetics* first appeared in 1834 in the writings of André Ampère, meaning the "science of government or control". But it was coined in its present meaning by Norbert Wiener, a mathematician who laid the conceptual foundations of the field, and his collaborators, in the 1940s. Wiener emphasized the deliberate blurring or eschewing of the natural-artificial distinction as a central tenet of his vision of cybernetics. The conviction was to discover unifying principles and a unifying language in which cognitive faculties as well as the behavior of things may be artifactually constructed and they could be explained with deeper theoretical insight. Developments in this field gave birth to modern "intelligent" robots. This strategy represents the first example of the radical shift in the study of mind: the shift from the investigation led by philosophers and conducted in nontechnical terms to a highly specialized form of research requiring the use of all the mighty resources of contemporary engineering (Franchi and Güzeldere 2005).

The original formulators of 'cybernetics' evidenced what can be called a 'mechanistic orientation' in two specific senses. The first sense involved prioritizing rigorous and non-vitalistic models for the system being studied. The second sense involved concentrating on the subject system's dynamics rather than the system as some static object of inspection. This orientation set the context for seeking an intersection among explanatory constructs for living and non-living systems. This notion of 'system' is fairly well circumscribed by its colloquial connotation of a discernible running entity which consisted of a finite set of constituent components. Trivially, this construct can be applied to anything constituted of parts, and with some dynamics. Non-trivial 'systems' are those in which the components actively participate in the constitution, configuration, and *behavior* of their collective set seen as a whole. The dynamics of these components' interactions, combined with the distinct set of interactions between their collective 'whole' and an environment, make for two domains of behaviors relevant to understanding that 'system'. This search entailed a belief that at some level of abstraction biological and non-biological systems could be addressed equally. It also involved a faith that at such a level of abstraction there were uniform and reliable principles to be discerned and analyzed. By the early 1940's such principles were being recognized as a result of implementing automatic behaviors in complex machines and emulating complex neural activities with machines.

The first computer systems exhibiting cognitive capacities considered "intelligent" (language understanding, learning, reasoning, problem solving) were presented in 1956, at the Dartmouth Conference. At the same time, John McCarthy coined the term *Artificial Intelligence*. McCarthy's main assumptions were that if a machine can do a job, then an automatic calculator can be programmed to simulate the machine. Knowledge was viewed as something that could be stored, coded, matched, and displayed. An artificial brain could be built simply by telling it what it needs to know. It was hoped that if we could represent the knowledge necessary to describe the world and the possible actions in a suitable formalism, then by coupling this world description with a powerful inference machine one could construct an artificial agent capable of planning and problem solving.

The turning point came with the development of knowledge-based systems in the 1960s and early 1970s. It has been explained as a "paradigm shift" in Computational Intelligence toward knowledge-based systems. The thousands of knowledge-based mechanisms or "expert systems" following it became visible demonstrations of the power of small amounts of knowledge to enable intelligent decision-making programs in numerous areas of importance. Although limited in scope, in part because of the effort to accumulate the requisite knowledge, their success in providing expert-level assistance reinforces the old adage that knowledge is power. But it was this advantage what at the end acted as their main handicap. Artificial Intelligence began to lose its holistic character towards a general theory of intelligence, acquiring a more "application" orientation. Knowledge-based systems were useful tools, although they do not simulate how humans think.

About the same time, computer specialists began to realize that scientific reasoning can be described in terms of problem solving search (Simon and Lea 1974, Langley et al. 1987, Klahr and Dunbar 1988, Klahr 2000, Kulkarni and Simon 1988, Thagard 1988, Valdés-Pérez 1995, 1996, Valdés-Pérez et al. 1993, Wagman 2000). According to that view, scientific theories may be considered as complex data structures in a computational system; they consist of highly organized packages of rules, concepts, and problem solutions. The idea is that scientific knowledge directs problem solving search through a space of theoretical concepts. This specific knowledge matches against different possible regularities in the data and take different actions depending on the kind of regularity the system has perceived among external data. Some of this knowledge proposes laws or hypotheses, others define a new theoretical term, and yet others alter the proposed scope of a law. Different data led to the application of alternative sequences of knowledge operators, and thus to different conclusions.

Generating a scientific explanation would be then a type of problem solving search, in which the initial state consists of some knowledge about a domain, and the goal state is a hypothesis that can account for some or all of that knowledge in a more concise form. A space of instances and a space of hypotheses should then be used, with the search in one space guided by information available in the other. That is to say, the use of instances constrains the search for hypothetical statements of the causal relationship. Hypotheses are evaluated through known instances of the causal relationship. In looking for appropriate instances of examples, scientists are faced with a problem solving task paralleling their search for hypotheses. They must be able to plan by making predictions about which observational (or experimental) results could support or reject various hypotheses. This involves search in a space of data that is only partially defined at the outset. Constraints on the search must be added during the problem solving process. This, in turn, requires domain-general knowledge about the pragmatic constraints of the particular discovery context.

If a computer program could be developed to *do* what we usually call "science", then an intelligent robot able to substitute us in the tedious task of studying ourselves should be also possible. By 1965, Herbert Simon predicted that machines will be capable of doing any work a man can do by 1985. When that date arrived, and the promised intelligent machines were still in the dreams (and nightmares) of programmers and computer hardware makers a critical approach to the very idea of thinking machines began. This failure precipitated the separation and rivalry of the two founding disciplines: Cybernetics and Artificial Intelligence. Cybernetics fundamental ambition was to produce a physically touchable theory of that most unphysical entity: the mind itself. The cybernetics researchers began their investigation of nervous systems by creating automata creatures reproducing what we (animals) can *do*. The artificial intelligence community ignored this approach in their early work and instead set the sights directly on the "intellectual" side of human thought, in experiments running on large stationary computers dedicated to mechanize pure reasoning.

As its history proves, computational intelligence is not just about robots. It is also about understanding the nature of intelligent thought and action using computers as experimental devices. In the following pages I will consider whether this is also a possible "metaphor" to understand the way archaeologists think.

3 Archaeological reasoning as computation

The first requisite for an automated archaeology should be based on the dream of Hobbes, "rationalization as computation". Formal logics, mathematics and computers have been used in archaeology, but their vast majority of archaeological applications pertain to the domain of methodology or even worst to the design of data collections. In many cases, such efforts were not directed to the examination of the structures of archaeological reasoning and argumentation.

Although computers and statistics began to be used in archaeology since the 1950s, we should wait until the end of the 1960s, when "new archaeologists" began to explore Hobbes argument. In so doing they made emphasis on the need to make disciplinary assumptions formal and explicit. Such authors considered that computer methodology provided an expanding armory of analog and digital techniques for computation, experimentation, simulation, visual display and graphic analysis. In that sense, it fulfilled a second requirement for automated archaeology: Russell's challenge for eliminating the subjectivity of perception and explanation in terms of some kind of externalized *demon*. Mathematical techniques, as sense-extending machine tools could either be used like the microscope to examine the fine structure of low-level entities and processes in minute detail, or like the telescope to scrutinize massive ensembles over vast scales. They also seemed to provide powerful hammer-and-anvil procedures to beat out archaeological theory from intransigent data - thus on one hand these methods can be used to construct models and simulate their consequences over a range of states, identifying test-conditions - on the other hand the computer may be used to analyze and test real data and measure their expectations under the model against the reality.

Doran and Hodson (1975, p. 74, see also Doran 1970) already suggested that the question "Can a computer do this?" is almost always rephrased as "Can this procedure exactly be specified?". In his book, *Analytical Archaeology* (Clarke 1968: 512-513) David Clarke noted three ways to explore rationalization as computation in archaeology:

1. using descriptive statistics for concept definition and quantification
2. using analytical inductive statistics to handle relationship concepts
3. using isomorphic systems of symbols arranged in axiomatic schemes, models or calculi to handle the regularities in complex data

The main problem in those years was that many scholars regarded mathematics and statistics as an analytical tool to be "used", and not as a way to transform rationalization into computation. Only some delimited aspects of archaeological reasoning were computationally formalized, like classification and seriation. Some emphasis was placed on statistical hypothesis testing, but there was very poor application of mathematical formalism to the theoretical issues of archaeology, despite recognition of the value of axiomatically or formally expressed theory (Read 1990). The naive use of Hempelian hypothetico-deductive reasoning mechanism as a "method" to test hypotheses is a good example. Statistics was placed at some part of the reasoning cycle, leaving the rest of the explanatory process in traditional narrative terms. A cautious note by Doran and Hodson (1975), the founding fathers of quantitative archaeology, is very interesting in this regard. They found the claims for a "formalized" approach to archaeology greatly exaggerated and therefore dangerous. While they share some of the dissatisfaction with subjectivities in archaeological explanation at that time, the proposed solution –the hypothetical-deductive method- was considered as a bizarre mixture of naivety and dogmatism. Formalization was still regarded as an "alien" conception.

A similar theme was iterated by Cowgill (1986: 369) in a review article titled, *Archaeological Applications of Mathematical and Formal methods*. There he referred to three broad categories comprised of "archaeological observations, analytical methods, and socio-cultural theory", but then observed that although some anthropological reasoning was expressed directly in computational (mathematical) terms, most of it was still expressed in a subjective narrative way.

The first synthetic models of archaeological inference were proposed by D. Clarke (1972), M.B. Schiffer (1976) , M. Borillo (1977), among others. They argued for the formalization of the acquisition of archaeological knowledge in terms of sets of laws, correlates and cultural and natural transformation processes. The cybernetic theory of the 50s and 60s provided the language necessary for that formalization. Instead of considering "archaeology" as a machine, "new archaeologists" regarded human society, and even the human individual (but not the archaeologist!) as a machine, forming a complex whole or "system". Here, "machines", "automata", and "societies" were synonymous. Archaeologists considered they should study the relationships between "components" to discover how the system worked in the past. The links between elements or subsystems were examined in terms of *correlational structure* (Clarke 1968).

Only in the middle 1980s, a step forward towards a full formalization was made, when archaeologists realized the need of imposing a concordance between the language of the model, the assumptions of the model and its interpretability (Carr 1985, Read 1985). The problem arising with axiomatization was not whether archaeologists have developed theory re-castable in axiomatic fashion but whether there are principles or relationships suitable for restatement as axioms for an axiomatic construction. The real fact is that archaeologists do not know exactly what archaeology is.

4 Simulating archaeologists

Jean Claude Gardin arrived by accident to professional archaeology at the end of the 1950s, and consequently he always had a rather external view of what archaeologists do. Instead of a normativist approach to archaeology, suggesting the best way of *constructing* the archaeology we need, he took an analytic point of view, looking for ways of *deconstructing* what archaeologists *believe they do.* His purpose has been to expose the logical flaws in argumentation and so to improve the logical execution of reasoning. This would allow the study of archaeological logic itself.

According to Gardin, the concrete expression of reasoning in any dominion of science is the text where the author has expressed the mental operations that have lead to him/her from the observation of certain empirical facts, to the affirmation of certain explanatory proposals. This methodology looks for the necessary bridges between facts and theses and the links between explanations. It has been called *logicist analysis* (Gardin et al. 1981). Its goal is to reduce the content of the text in its main components, studying their fundamental connections. The schematization of an archaeological paper is not an abstract or a summary of the paper, but a reformulation of its content in a condensed form. Gardin uses the word "condensation" as in physics: a rearrangement of something into a more compact volume, without loss of substance. He and his colleagues "have deconstructed" numerous scientific works (mainly archaeological) in this way. This approach is precisely a framework for analyzing and modeling the questions and answers that bracket a scientific text, and there is an obvious intuitive link between meaning, questions, and answers.

Gardin assumed that our theoretical constructs can be expressed in terms of a "calculus". Archaeological theories can be formulated as computational structures with two components. The first one is a *facts base*, here understood as a set of declarative propositions that include not only descriptions of archaeological materials and their context, with associated archaeometric data, but also a large number of referential statements. Those statements are not usually regarded as "data"; they include primarily vast sets of analogies, "common sense", shared belief, ideologies, etc. The second component is an inferential tree made up of rewrite operations, which reproduce the chain of inferences from the archaeological record ("facts", represented as P_o) to different explanatory statements (P_n). Between the extremes of the argumentation, there are intermediate theses (*Pi*). Scientific reasoning builds chains of oriented propositions P_o, P_1, P_2..., P_n in terms of successive operations P_i --> P_{i+1} (Gardin 1980, 1991, 1993, 1994, 1998, 2003).

"Rules" are the key; not laws, which are inviolate, but rules that can be changed and indeed are always changing in a reflexive relationship allowing the expert (human or machine) to accommodate new information. Given some empirical data (observations) about a particular archaeological case, and some bit of associative knowledge (If...Then) (hypotheses and interpretations considered valid in a Social, Anthropological or Historical Theory), the archaeological problem can be explained in terms of the knowledge stored in a series of rules. In other words, given some visual input and a candidate explanatory causal model, a correspondence can be established between them. This means that a small number of features are identified as matching features in the input and the model. Based on the corresponding features, a decision rule linking visual features with their causal process (social activity) is uniquely determined. The recovered decision rule is then applied to the model. Based on the degree of match, the candidate causal event is selected or rejected. To be accepted, the match must be sufficiently close, and better than that of competing solutions.

The rules discovered by logicist analysis may be subjective, but they are explicit. Anyone can produce the same results, so that although the system is subjective, it will be consistent when different subjectivities (i.e. different individuals) use it. The acceptance of the assumptions on which the problem solution is based leads to consistency, and direct comparability between results produced by different people; this fulfills the basic requirements of objective data within the consensus reality of mutual users of the program. Therefore, logicist analysis can extract objective-like knowledge, but the complexity of the dynamic process is retained and the data is produced in the form of probabilities that can be compared as if they are objective data within a defined consensus reality.

Analogies between logicist analysis and some aspects of artificial intelligence are patent, although both representation schemas evolved in parallel without further implications (Gardin 1980:123-125, 1991, Gardin et al. 1987). Formal characteristics of Expert Systems technology appear to be very similar to the general structure of logicist analysis rewrite rules. The "deconstruction" of a scientific text in terms of rewriting operations agrees with the "extraction" of the expert knowledge in terms of production rules. In the same way that the knowledge engineer tries to find out how a human expert thinks before introducing "prior knowledge" inside the computer program, logicist analyst tries to study what is hidden inside a scientific text written in natural language. Similar approaches have been those by Borillo (1984), Stutt (1989), Segal (1994), Winder (1996), Orlandi (1997, 2002), Tsaganou et al. (2003), Zhang et al. (2002).

Gardin accepted that the way archaeologists take decisions can be mechanized. Although he never tried to build an automated archaeologist, his suggestions moved some archaeologists to create what at first look seem to be "automated archaeologists". The most obvious application of this "automatization" of archaeological reasoning is the domain of archaeological typologies. In the same way, the function and chronology of ancient buildings can be correctly explained from their observed architectural features, and the visual characteristics of human and animal bones can be used to recognize them as instances of well defined explanatory categories. It is also possible to mechanize the process of microscope samples classification for ancient wood taxonomy determination. Some other systems help scientist to decode decorative patterns in pottery or rock-art. Other archaeological applications have explored the possibilities of whole artifact identification from the perception of sherds. Applications of automated problem solving methods do not finish here. An expert system can be programmed to help archaeologists to interpret the results of archaeometric analyses, within the framework of provenance studies. Such a system would produce one (or several) "diagnoses" according to the geographic origin of raw material, from a database of analyzed samples of known origin provided by the user. See a review of such applications in Barceló (2008).

5 What Computers Could not Do Years Ago and What They Actually Do

Herbert Simon prediction that machines will be capable of doing any work a human can do by 1985 was soon considered over-optimistic for some authors, exaggerated for others, or even wrong for many computer scientists and philosophers. Some years before that landmark date it became clear that intelligent machines could not be produced. In fact, not even today, 25 years after the deadline, we have arrived at a true computational intelligence. In the same way, and around the same years Gibbon (1984: 383), though espousing the value of formal and axiomatically expressed theory in archaeological reasoning, bluntly commented that "no theory within archaeology has ever been formalized". The impossibilities of machine intelligence and automated archaeology seem to have been detected simultaneously: there is no easy way to translate rationalization into computation.

In any case, we have to accept that the very idea of "rationalization as computation" never found the place in archaeology (nor in any other social science) it merited. The most promissory computational techniques of those early days were accused of excessive simplification, of forcing knowledge, or distorting it, and of failing to exploit fully the knowledge of the expert (Hugget and Baker 1986, Wilcock 1986, 1990, Doran 1988, Gallay 1989, Lagrange 1989, Shennan and Stutt 1989, Francfort 1990, Puyol-Gruart 1999).

However, there is nothing suspicious in the approach. The success of expert systems in parallel disciplines is very evident if we consider the thousands of references (Liao 2003, 2005), and it is due to their working within a world in which the range of meaning for terms is circumscribed within a carefully selected micro-world. Yes, may be they are not a model of human reasoning, as considered by Gardin, but this technology really works! The problem is that archaeology has not yet arrived to a relevant degree of formalization, given absurd prejudices and the weight of individual authority. Robots are not guilty, but humans that have not learn how to program them!

In archaeology, the so called "radical critique" of the 1980s, with its comic mixture of hermeneuticism, individualism, humanism, and stupidity deformed the debate when it regarded archaeology as literature. There are still scholars considering that any archaeological analysis is a mere text product of an individual writer. Consequently the explanation of past behavior has the same value as a literary product. Even the practice of archaeology can itself be reduced to "theatre". Given that robots can not act, there are no automated archaeologists! Given that rationalization is seen as art (literature), it is considered that it cannot be rendered computable, because the act of literary creation is intrinsically incomputable.

The debate never arrived to such extreme in the Artificial intelligence debate. Although by 1985, computer scientist and cognitive psychologists were well aware that no general theory of rationalization could be rendered computable, and no "artificial human brain" has ever been programmed, they had already proved that intelligence could be mechanized in very restricted domains. Within the last two decades, the view of computational intelligence based on pre-fixed plans and searching in restricted knowledge-bases using well-defined operators for activating already existing sequence of explanations (i.e. expert systems) has come under scrutiny from both philosophers and computer scientists.

The main consequence of this deep criticism was the renaissance of the cybernetic approach in the late 1980s, and its integration with new paradigms of cognitive science, philosophy and a so called "new" artificial intelligent paradigm. A shift in perspective from *knowledge as stored* artifact to knowledge as *constructed capability-in-action* inspired a new generation of cyberneticists in the fields of situated robotics (Winograd and Flores 1986; Brooks, 1989, 1991, 1999; Brooks et al. 1998, 1999; Clark, 1993, 1997; Franklin 1995; Hendriks-Jansen 1996, Clancey 1997; Engel and König 1998; Pfeiffer and Scheier 1999; Anderson 2003; Iyida et al. 2004). To be intelligent, an intelligent machine should focus on the outside world, how this world constrains and guides its explanatory behavior. The automated system we would like to build is the agent-in-the-right-context, an agent constructing descriptions by adapting old ways of perceiving, by putting models out into the world as artifacts to manipulate and rearrange, and by perceiving generated descriptions over time, relating them to past experiences or future consequences.

Machine Learning appears then as the key word in the New Cybernetics. That is to say, we do not simply ask: "What knowledge structures should be placed on the head of a robot able to do archaeology?" Instead of storing declarative sentences in computer's memory, we should build a machine able to *learn* from its own explanations and mistakes. If we want to go beyond the traditional expert-system approach, we should make emphasis not on database consultation, analogy, and simple statistical decision-making, but on learning and categorizing, and on how meaning can be generalized from known examples of a given concept. Fortunately, learning is not an impossible task for computers. New generation adaptive algorithms (neural networks, support vector machines, genetic algorithms) appear to be formally true universal inductive algorithms, and they can be used to solve many archaeological problems (see, Barceló 2008).

Programming computers to be able to solve most learning problems is a cross between statistics and computer science. The idea is to program a system able to look for common features between positive examples of an observed or simulated causal relationship to be predicted, and common differences between its negative examples. In contrast with discrete Aristotelian logics, machine learning models provide more *graded* answers to archaeological problems. Such programs integrate information from a large number of different input sources, producing a continuous, real valued number that represents something like the relative *strength* of these inputs. These graded signals can convey something like the *probability* of the answer or explanation in some specifically constrained circumstances.

Computer scientists are intensively exploring this subject and there are many new mechanisms and technologies for knowledge expansion through iterative and recursive revision. Artificial Intelligence offers us powerful methods and techniques to bring about this new task. Fuzzy logic, rough sets, genetic algorithms, neural networks, Bayesian models and agent-based systems are among the directions we have to explore. These paradigms, differ from usual methods in that they are (in comparison at least) robust in the presence of noise, flexible as to the statistical types that can be combined, able to work with feature (attribute) spaces of very high dimensionality, they can be based on non-linear and non monotonic assumptions, they require less training data, and make fewer prior assumptions about data distributions and model parameters. The huge number of learning algorithms and data mining tools make impossible that we can review the entire field in a single paper (see Jones 2007, Luger 2008, Munakata 2008, Hassanien et al. 2009, Bar-Cohen et al. 2009)

No aspect of this discussion has entered in the archaeological debates of our time. Critics of the "rationalization as computation" view of archaeological discipline are ignorant of this renaissance of the cybernetic paradigm, and its integration with new paradigms of cognitive science, philosophy and new programming approaches. What at the beginning seemed correct criticisms of the view of human society as a *simple* machine, soon became an hysterical rejection of formalization and any possible surrogates: computers, statistics and formal logics (Shanks and Tilley 1987). The idea of "art" or "humanities" has been violently vindicated favoring explanation via "common sense", ignoring the fact that artificial intelligence is technologically achievable provided we change the classical approach of its early days: if we want to reproduce human intelligence in a machine, we should make emphasis on three central aspects: development, interaction, and integration. Development forms the framework by which machines should imitate the way humans successfully acquire increasingly more complex skills and competencies. Interaction should allow robots to use the world itself as a tool for organizing and manipulating knowledge, it allows them to exploit humans for assistance, teaching, and knowledge. Integration should permit an automated archaeologist to maximize the efficacy and accuracy of complementary mechanisms for perceiving and acting.

Therefore, what would give a more "intelligent" character to automata applications in the archaeological domain will not be a passive storing of individual rules, but an enhanced ability to learn and to react in a certain way to a certain stimulus. If we want to go beyond the usual archaeological explanations based on template matching, we should make emphasis not on database consultation, analogy, and decision-making, but on learning and categorizing, and on how meaning can be generalized from known examples of a given concept. That is, the automated archaeologist should develop its own cognitive machinery (what it knows) as opposed to construct a data structure on which a preexisting machinery operates.

If we build an "intelligent" robot based on those aspects, we will obtain a machine, which is not born with a complete reasoning mechanism. Instead, it should undergo a process of development where it will perform incrementally more difficult tasks in more complex environments en route to an advanced state. In other words, our robot should be capable of improving its capabilities in a continuous process of acquiring new knowledge and skills, and reacting to unusual events such as incomplete input, lack of prior knowledge. To fulfill this requirement, it is suggested that an automated archaeologist should not be fully programmed since the beginning, but it be able to learn as soon as available data changes in quality or in quantity. The gradual acquisition of interpretive skills and the consequent gradual expansion of the automated archaeologist capacities to explain archaeological observables (creating more and more self-training data as it does so) will define then the cognitive behavior of our machine. This strategy facilitates learning both by providing a structured decomposition of skills and by gradually increasing the complexity of the task to match the competency of the system. Behaviors and learned skills that have already been mastered prepare and enable the acquisition of more advanced explanations by providing subskills and knowledge that can be re-used, by placing simplifying constraints on the acquisition, and by minimizing new information that must be acquired.

Consequently, an automated archaeologist would not be obliged to apply discrete responses to discrete facts. What an intelligent machine should "see" is a probabilistic manifestation of its past "experiences" with similar inputs. Percepts are neither correct nor incorrect representations of reality but simply a consequence of having incorporated into the explanatory mechanism the statistics of explanatory success or failure. Of course, this utilitarian-cum-probabilistic approach to perception does not imply that the mechanisms of perceptual problem solving are completely chaotic and unlawful in character.

The failure of the Artificial Intelligence early prospect was credited to a view of intelligence as an abstract machine. In the same way, the failure of the New Archaeology of the 60s was its insistence on simple universal theories of human behavior. The reaction should be based on a move towards a view of knowledge as something created as transformations of previous knowledge. Information does not exist in the world waiting to be extracted by a robot, but, rather, it should be situated in meaningful contexts. Perceiving a world implies distinguishing "possibilities for action" and not naming or identifying *per se*. Explanation cannot be properly understood, if considered independently of the context in which it occurs. The historical, cultural, and social context of the interactions of an intelligent machine is crucial to the understanding of the ongoing process. That is to say, the archaeological record is here defined in terms of the recognition of the circumstances to act with or upon (explanation). Being a perceiver, an intelligent machine should literally create a phenomenal world, because the process of perception first defines relevant distinctions in the sensory environment. Universally intelligent machines cannot simply be programmed with knowledge, since that would imply a designer-based high-level ontology that would in turn lead to a system incapable of interacting efficiently with the world on its own. A key implication is that a supposed intelligent machine would have to interact with its environment on its own, because solutions to any problem do not come from nowhere. The admittedly strange idea that what an automated archaeologist "sees" is a probabilistic manifestation of its past "experiences" with similar inputs, rather than a logical analysis of what it perceives in this actual case may be difficult to accept. The archaeological record is then neither correct nor incorrect representations of reality but simply a consequence of having incorporated into visual processing the statistics of visual success or failure. Of course, this utilitarian-cum-probabilistic approach to perception does not imply that the mechanisms of perceptual problem solving are completely chaotic and unlawful in character. That is to say, archaeological observables should be understood as emergent properties of sensory fields, not static things in the environment that are merely detected, selected, or picked up. In this conception of perceptual problem solving, the automated archaeologist should see its empirical significance, or more formally, the probability distribution of the possible sources of the stimulus, in response to any given stimulus. Understanding what a robot can see and why will depend on understanding the probabilistic relationship between stimuli and their sources during the automated archaeologist past experience. As a result, the percepts that are entertained would accord with the accumulated experience of what the visual and non-visual inputs in question had typically signified in the history of this individual automated archaeologist.

The approach exposed here challenges the received picture of an archaeological explanation as an invariant structure. Solving archaeological problems is an activity. We have to change the way we understand explanatory concepts. They are not verbal labels we attach to some percepts by means of a previously existing rule but a cognitive action, or a requisite to a next action. Explanations should be based on purposeful, goal-directed mechanisms emerging from a dynamical system that has been calibrated by learning (trial and error, experimentation, analogy) to make the right choices in the proper circumstances. What I am suggesting is that when explaining, our automated archaeologist conceptually navigates in a potential field of explanations looking for attractors (goals) and repulsions (constraints). Upon detecting the goal, the explanation moves toward it, executes it and then follows until another goal or constraint is found. It repeats this sequence of actions until it has returned all attractors in the potential field. Since the robot does not manipulate propositions, any account of automated explanation that would draw on connectionist principles would not be able to limit itself to principles of logical inference in describing how some belief was arrived at. Rather, it is necessary to rely on something like the notion of maximal satisfaction of soft constraints to describe how the a machine behaves cognitively, and in *evaluating* its performance we would presumably consider whether the constraints it satisfied in arriving at its output state were the appropriate constraints. This would lead us into an evaluation of how an automated archaeologist has learnt, specifically, whether its training had resulted in ways that enabled it to respond to inputs in a manner that was most likely to meet its needs in the environment. This would constitute a major change, since epistemology has generally been pursued through conceptual analysis, not empirical inquiry (Bechtel and Abrahamsen 2005).

6 Towards a Computational Philosophy of Archaeology

Don't panic! I am not arguing that an artificial archaeologist will substitute human archaeologists, because it works better and cheaper than us. We all know that Artificial Intelligence will eventually produce robots whose behavior may seem dazzling, but it will not produce robotic persons (Bringsjord 1992). Automatic archaeologists will DO a lot, but they won't BE a lot. In so saying, I am following Roger Penrose, who notes in his book, *The Emperor's New Mind*: "it would be fair to say that although many clever things have indeed been done, the simulation of anything that could pass for genuine intelligence is yet a long way off" (Penrose 1989). Computational mechanisms cannot by themselves carry the weight of a scientific explanation. No machine is ever likely to provide an adequate explanatory analogy for the human brain or mind. Machines will not produce "for free" the categories we need for explaining social action in the past. "Real" machines are too simple and limited in their functions, and a "virtual" machine like our automated archaeologist merely imports formal redescriptions of the surface phenomena into the explanatory model. Nevertheless, computer-based models mimic human behavior, and therefore they are good models of cognitive behavior rather than models of brains or minds.

In so saying, I am not arguing that artificial archaeologists (computer programs) run like human brains or that computer representations should be isomorphic to "mental" states. I am not pretending to simulate any of us, when we are doing archaeology, but to create something different. I make emphasis on the necessity of understanding mental processes by making evident the underlying abstract causal nature behind cognitive tasks. This structure can be understood objectively and duplicated in computers. If a computer can be programmed to perform human-like tasks it offers a "model" of the human activity that is less open to argument than the empirical explanations that are normal in philosophy. The purpose is to understand how intelligent behavior in archaeology is possible. I suggest the possibility of building a "virtual" automatic archaeologist in such a way that the shortcomings of *natural* archaeology be avoided.

It is easy to see that the purpose of an *Android epistemology* (see Ford et al. 1995 for the meaning of the term) is not to study machines in themselves, but *human cognition*. Everybody is aware of the enormous complexity of a human being. If we look at something comparatively simple like a robot, we may be surprised how complicated it is, all the topics and issues one has to think about, design, and build before anything goes. Therefore, I am not arguing that "natural" archaeology can be fully described in terms of an "automatic" archaeology. I am suggesting that computational cognitive models of archaeological abilities should be based on the study of particular human capabilities and how humans solve certain tasks, but such models never will be *like* human archaeologists, nor do I pretend to substitute human scientific endeavor by slave androids.

Such an approach can be characterized as "understanding by building" (Braitenberg 1984, Rutkowska 1993, Adams et al. 2000, Holland and McFarland 2001, Fernandez 2003, Dawson 2004, Drennan 2005, Doyle 2006). Computer hardware and programming techniques enable the model builder to construct virtual creatures that behave in intelligent and flexible ways under natural conditions. They provide powerful (and perhaps indispensable) tools for building such creatures, but they can play no role as explanatory kinds. The hardware and software used in such models carry no explanatory burden.

A synthetic approach to archaeological reasoning implies building computer simulations that instantiate detailed theories of archaeological explanation. Therefore, it is closely coupled with the notion of design. Although we normally think about design as an activity for engineers, the design perspective has proven extremely fruitful in cognitive science for studying natural intelligence. In a synthetic methodology the key question is translated into a design issue: How can we design an *agent* so that it will exhibit the desired behaviors? Making a computational model forces us to be explicit about exactly how the relevant process actually works below the level of consciousness. Such explicitness carries with it many potential advantages. As such, it should provide novel sources of insight into archaeologist's behavior. It should allow a way to deal with complexity in ways that usual verbal arguments cannot, producing satisfactory explanations of what would otherwise just be vague hand-wavy arguments. Explicitness can contribute to a greater appreciation of the complexities of otherwise seemingly simple process. Furthermore, by integrating cognitive development into robot engineering a system comes to know new things about its environment, integrates those with what it already knows, and utilizes that knowledge.

The present approach owes much to the work of David Marr on computational psychology (see Marr 1982). According to his suggestion, theories of computation –and an automatic archaeology is just a theory of computation- are at the top of a three-level hierarchy. At the highest, most abstract level, there is the fundamental problem we want to solve. An intermediate level describes one solution to that problem, particularly in terms of the representations and operations required for the solution. The lowest description is concerned with the physical implementation of the algorithm to generate the solution.

The top level can be equated with the *knowledge-level* (Newell 1982), where a number of predisposed cognitive actions are specified. The knowledge level is a level of abstractions where we can consider what the problem solver knows, without necessarily knowing anything about how to reason. The components are goals and actions. You should imagine this level as containing terms which are assumed to correspond to the "knowledge" archaeologists deal with. It corresponds to a theorem in mathematics: Given a specific goal, specific input to work with, and specific "assumptions" about the world, one may hope that there often might be only one set of computations that would allow a system to produce the required output. In other words, one assumes that it is possible to identify the "optimal" computation or function performed by any archaeologist in a given context, and that whatever the archaeologist is doing, it must somehow be accomplishing this same optimal computation. Under this view, it doesn't really matter how the archaeologist perform his/her task, because it is ultimately driven by the optimality criterion of matching expected demands for items, which in turn is assumed to follow general laws. The additional difficulty is that, optimality can rarely be defined in purely "objective" terms, and so often what is optimal in a given situation depends on the detailed circumstances.

For this reason, if the first level is a theory of what should be computed, the next level down is a theory of the algorithm, which specifies *how* a computation should be performed, and specifies the conditions where the procedure can generate valid results. The theory of the algorithm specifies an explicit set of steps that will guarantee a given output when provided with a given input. However, such an intermediate level describes the performed function in terms abstracted from the details of the system's physical implementation (the human brain or the mechanic hardware circuitry), and typically, many different algorithms can carry out a computation, so we need a lower level, where all the actions are described without giving an explicit account of what overall function is performed. Note that while the higher levels are about what the agent –human or machine- believes about the external world and what its goals are in terms of the outside world), the lower levels are about what goes inside an agent in order to reason about the external world (Poole et al. 1998, p. 180).

Consequently, whether a theory of computation can be seen as a description of the problem, a theory of the algorithm is a description of a particular solution. The *knowledge* level, or the theory of the computation may be treated as paramount because it characterizes the problem that must be solved by a system –and until one understands what a system does in specific circumstances, one cannot specify the details of how the system works. We can place the algorithm at a lower level of the hierarchy because it does not characterize a problem, but rather specifies a possible solution to a problem that is characterized at the more abstract level.

The very idea of a *Computational Philosophy of Science* has been called "the most self-contradictory enterprise in philosophy since business ethics". I'm using a direct reference to Paul Thagard's book "Computational Philosophy of Science" (1988), where he presented a challenge to the philosophical community: philosophical theories of scientific method, if they are worth their salt, should be represented as computer programs. Computational philosophy of science is a kind of experimental epistemology. It investigates methods for representing knowledge and for modeling reasoning strategies that can manipulate that knowledge. Once one has a working system, one can experiment with "what if" scenarios, adding or removing strategies to determine their effects (Shrager and Langley 1990, Stary and Peschl 1995, Peschl 1996, De Jong and Rip 1997, Darden 1997, Fernández 2003, Alai 2004, Humphreys 2004, Magnani 2004, Magnani and Dossena 2006).

In other scientific domains, the performance of humans at a particular task has been used to design a robot that can do the same task in the same manner (and as well) (Moravec 2000, Nolfi and Floreano 2000, Bryant et al. 2001, Murphy 2002, Florian 2002, Santore and Shapiro 2004, Kovacs and Ueno 2005, Trafton et al., 2004, King et al. 2004, Tamburrini and Datteri 2005, Datteri and Tamburrini 2006). It has been shown how 'robot scientists' can interpret experiments without any human help. Such robots generate a set of hypotheses from what it is known about a scientific domain, and then design experiments to test them. That is, a robot scientist can formulate theories, carry out experiments, and interpret results. For instance, the robot biochemist developed by Ross King and his colleagues, does everything a flesh-and-blood scientist does—or, rather, it does what philosophers of science say that scientists ought to do. That is, it can infer hypotheses to explain observations, infer experiments that will discriminate between these hypotheses, actually do the experiments, and understand the results (Bryant et al. 2001, King et al. 2004).

Consequently, the design of an automated archaeologist should not be considered a mere science fiction tale. It is a technological reality. Research in cognitive robotics is concerned with endowing robots and software agents with higher level cognitive functions that enable them to reason, act and perceive in changing, incompletely known, and unpredictable environments. Such robots must, for example, be able to reason about goals, actions, when to perceive and what to look for, the cognitive states of other agents, time, collaborative task execution, etc. In short, cognitive robotics is concerned with integrating reasoning, perception, and action within a uniform theoretical and implementation framework. The question of whether it is possible to such machines to automate the scientific process should be of both great theoretical interest and increasing practical importance because, in many scientific areas, data are being generated much faster than they can be effectively analyzed.

Intelligent robots are here, around us. I've never heard of a claim against wash machines selecting "intelligently" the best way to wash a specific tissue, or a photo camera with an "intelligent" device measuring luminance and deciding by itself the parameters to take the picture. So, why do we fear of a machine classifying a prehistoric tool and deciding "intelligently" its origin, function, and/or chronology? The so-called "intelligent" machines incite instinctive fear and anger by resembling ancestral threats -a rival for our social position as more or less respected specialists. Nevertheless, it is important to remember what I said at the beginning of the paper, that any formal definition of an *automaton* includes both human beings and the "intelligent" robots within the same reference class. People are indeed mechanical, in both mind and body, but are not necessarily machines or material machines (Doyle 2006).

Nevertheless, the real question is not whether machines think but whether archaeologists do.

7 Conclusions

The history of Computational Intelligence is a history of fantasies, possibilities, demonstrations, and promise. This basic way of thinking of human beings as entirely material organisms construed in biomechanical terms still carries some shock value (see Pollock 1989, Moravec 2000, Kurzweil 2000, Hall 2007) and is highly contested. It is by no means a generally accepted view in philosophy of mind that the mental is reducible to the physical.

My view is exactly the opposite. I am taking as granted the fact that archaeological reasoning is analyzable. The basic idea behind it is simply that expertise, which is the vast body of task-specific knowledge, is transferred from a human to a computer. This knowledge is then stored in the computer and users call upon the computer for specific advice as needed. The computer can make inferences and arrive at a specific conclusion. Then like a human consultant, it gives advices and explains, if necessary, the logic behind the advice. Such a strategy has been successfully applied in many diverse domains as finance, medicine, engineering, legal studies, ecology and biological taxonomy (Liao 2003, 2005). Why not in archaeology? In these *post-modern* days, it may seem a provocation. Some readers may think that archaeology is just what archaeologists do, and there is no formal basis except the personal subjectivities of the archaeologist. The approach adopted here has a distinctive *processualist* flavor: there is a reality out here, and it can be analyzed in terms of the causal processes having generated it. However, recent advances in cybernetics and artificial intelligence show that there is no way to program an automated archaeologist, without taking into account some of the classical topics of *post-processualist* criticism, notably, the nature of context, goals, intentions, and motivations.

My personal approach is based on a fact that archaeologist couldn't evaluate 15 years ago: Computer programs do work in real science, not only in archaeology. Maybe they are more successful in other "harder" sciences, but we cannot deduce from this fact that Archaeology *is a different* kind of science. We should instead rebuild archaeology. Simulating or reproducing the way archaeologists think today is not the guide to understand archaeology, because *we are doing archaeology in the wrong way!* Computable archaeology –if you do not like the expression "automatic archaeology" is the proper way of exploring new ways of thinking old concepts.

The idea of an intelligent robot has been here presented as a model of archaeologists' behavior rather than models of their brain or mind. The discussion is between what is considered an *artificial* way of reasoning (computer programs) and a *natural* way of reasoning (verbal narrative). Critics of computationalism insist that we should not confound scientific statements with predicate logic operations, since discursive practices or argumentations observed in a scientific text are not "formal". By that reason, they are tributary, to a certain extent, from the Natural Language and the narrative structure (literary) of which scientific texts derive. I take the opposite approach: scientific problem solving stems from the acquisition of knowledge from a specific environment, the manipulation of such knowledge, and the intervention in the real world with the manipulated knowledge. The more exhaustive and better structured the knowledge base, the more it emulates a Scientific Theory and the easier will be the solution to the scientific problem, and more adequate the interpretations we get.

Acknowledgements

Thanks to my students at the Dept. of Prehistory (Universitat Autònoma de Barcelona, Spain) who are beginning to explore the many possibilities of artificial intelligence in archaeology. Parts of this research have been funded by the Spanish Ministry of Science and Innovation, The Generalitat de Catalunya and the Universitat Autònoma de Barcelona. More information about this project can be found in **http://prehistoria.uab.cat/Barcelo/IGIBook.html**.

Bibliography

Adams, B., Breazeal, C., Brooks, R.A, Scassellati, B., 2000, "Humanoid Robots: A New Kind Of Tool" *IEEEIntelligent Systems And Their Applications: Special Issue On Humanoid Robotics*, Vol. 15, No. 4, July/August 2000, Pp. 25--31.

Alai, M., 2004, "A.I., Scientific Discovery And Realism", *Minds And Machines* 14: 21–42, 2004.

Aleksander,I., Morton,H., 1993, *Neurons And Symbols. The Stuff That Mind Is Made Of.* London, Chapman And Hall.

Anderson, M.L., 2003, "Embodied Cognition: A Field Guide". *Artificial Intelligence*, 149, Pp. 91-130.

Arkin, R.C., 1998, *Behavior-Based Robotics*. Combridge

(Ma), The MIT Press.

Barceló, J.A., 2008, *Computational Intelligence in Archaeology.* Henshey (NY), Information Reference Source, IGI Group Inc.

Bar-Cohen, Y., Hanson, D., Marom, A., 2009 *The Coming Robot Revolution: Expectations And Fears About Emerging Intelligent, Humanlike Machines* Berlin, Springer.

Bechtel, W. And Abrahamsen, A. 2005. Explanation: A Mechanistic Alternative. *Studies In History Of Philosophy Of The Biological And Biomedical Sciences.* Volume 36, Issue 2, Pages 421-441

Borillo, M., 1977, "Raisonner, Calculer", In, M. Borillo (Ed) *Raisonnement et Méthodes Mathematiques en Archéologie, pp,. 1-31.* Paris. Editions CNRS,

Borillo, M., 1984, *Informatique Pour Les Sciences De L'homme.* Bruxelles, P. Madarga Publ.

Braitenberg, V., 1984, *Vehicles: Experiments In Synthetic Psychology*, Cambridge (MA), The MIT Press.

Bringsjord, S., 1992, *What Robots Can And Can't Be.* Boston: Kluwer.

Brooks, R. A., 1989, "How To Build Complete Creatures Rather Than Isolated Cognitive Simulators", In K. Vanlehm (ed) *Architectures For Intelligence*, Hillsdale, NJ, Erlbaum,pp. 225–239.

Brooks, R. A. (1991), `Intelligence Without Representation', *Artificial Intelligence Journal* 47, 139-160.

Brooks,R., 1999, *Cambrian Intelligence: The Early History Of The New AI.* Cambridge (MA), The MIT Press.

Brooks, R.A., C. Breazeal (Ferrell), R. Irie, C. Kemp, M. Marjanovic, B. Scassellati And M. Williamson, 1998, "Alternate Essences Of Intelligence", *Proceedings Of The Fifteenth National Conference On Artificial Intelligence* (Aaai-98), Madison, Wisconsin, July 1998, Pp. 961--976.

Brooks, R.A., Cynthia Breazeal, Matthew Marjanovic, Brian Scassellati, Matthew Williamson, 1999, "The Cog Project: Building A Humanoid Robot". In C. Nehaniv (Ed) *Computation For Metaphors, Analogy, And Agents.*, Lecture Notes In Artificial Intelligence 1562. New York, Springer, 52–87, 1999.

Bryant, C. H., Muggleton, S. H., Oliver, S. G., Kell, D. B., Reiser, P., & King, R. D. (2001). Combining Inductive Logic Programming, Active Learning And Robotics To Discover The Function Of Genes. Electronic Transactions On Artificial Intelligence, 5, 1–36.

Carr, C., (Ed.) 1985, *For* Concordance *In* Archaeological *Analysis: Bridging Data Structure Quantitative Technique And Theory*. Boulder, Westview Press.

Clancey,W.J., 1997, *Situated Cognition: On Human Knowledge And Computer Representations,* Cambridge University Press.

Clark,A., 1989, *Microcognition: Philosophy, Cognitive Science, And Parallel Distributed Processing.* Cambridge, The Mit Press.

Clark,A., 1993, *Associative Engines. Connectionism, Concepts And Representational Change.* The Mit Press. Cambridge (Ma).

Clarke, D.L., 1968, *Analytic Archaeology.* London, Methuen

Clarke, D.L., (Ed.) 1972, *Models in Archaeology.* London, Duckworth

Cowgill, G.L., 1986, Archaeological Applications Of Mathematical And Formal Methods. In D.J. Meltzer, D.D. Fowler, J.A. Sabloff (Eds). *American Archaeology, Past And Future: A Celebration Of The Society For American Archaeology.* Washington. Smithsonian Institution Press.

Darden, L., 1997, "Anomaly-Driven Theory Redesign: Computational Philosophy Of Science Experiments," In T. W. Bynum & J. Moor (Eds.), *Digital Phoenix: How Computers Are Changing Philosophy.*Oxford: Blackwell.

Datteri, E., Tamburrini, G., 2006, "Bio-Robotic Experiments And Scientific Method". In *Computing, Philosophy And Cognition.* Edited By Magnani, L., Dossena, R. College Publications, London, Pp. 397-41

Dawson, M.R.W., 2004, *Minds And Machines. Connectionism And Psychological Modeling.* London, Blackwell Pub.

De Jong, H., Rip, A., 1997, "The Computer Revolution In Science: Steps Towards The Realization Of Computer-Supported Discovery Environments". *Artificial Intelligence* 91 Pp. 225-256

Donahue, J.W., Palmer, D.C., 1994, *Learning And Complex Behaviour*. Boston, Allyn And Bacon.

Doran J. E., 1970, "Archaeological Reasoning And Machine Reasoning". In *Archaeologie Et Calculateurs* Efited By J.C. Gardin. Paris, CNRS. Pp 57-67.

Doran, J.R., 1988, "Expert Systems And Archaeology: What Lies Ahead?" *Computer Applications In Archaeology* (Bar International Series, 393), Pp. 237-241.

Doran, J.R., Hodson, F., 1975, *Computers In Archaeology.* Edinburgh, Edinburgh University Press.

Doyle, J., 2006, *Extending Mechanics To Minds. The Mechanical Foundations Of Psychology And Eonomics.* Cambirdge University Press.

Drennan,M., 2005, "The Human Science Of Simulation: A Robust Hermeneutics For Artificial Societies" *Journal Of Artificial Societies And Social Simulation* Vol. 8, No. 1 <Http://Jasss.Soc.Surrey.Ac.Uk/8/1/3.Html>

Engel,A.K., König,P., 1998, "Paradigm Shifts In The Neurobiology Of Perception". In *Intelligence And Artificial Intelligence. An Interdisciplinary Debate.* Edited By U. Ratsch, M.M. Richter And I.-O. Stamatescu., Berlin, Springer Verlag.

Fernandez, J., 2003, Explanation By Computer Simulation In Cognitive Science, *Minds And Machines* 13: 269–284,

Florian, R.V., 2002, Why It Is Important To Build Robots Capable Of Doing Science In C. G. Prince, Y. Demiris, Y. Marom, H. Kozima, & C. Balkenius (Eds.), *Proceedings Of The Second International Workshop On Epigenetic Robotics: Modeling Cognitive Development In Robotic Systems.* Edinburgh, UK. Lund University Cognitive Studies 94, Pp. 27-34, 2002.

Ford, K.M., Glymour, C., Hayes,P,J., Eds, 1995, *Android Epistemology*. Menlo Park/Cambridge/London. AAAI Press /The Mit Press.

Francfort, H.P , 1990, «Modélisation De Raisonnements Interprétatifs En Archéologie A L'aide De Systèmes Experts: Conséquences D' Une Critique Des Fondements Des Inférences». In *Interpretation In The Humanities: Perspectives From Artificial Intelligence.* Edited By J.C. Gardin And R. Ennals. The British Library Publications. Library And Information Research Report, 71.

Franchi, S., Güzeldere, G., (Eds.), 2005, *Mechanical Bodies, Computational Minds*. Cambridge (Ma), The Mit Press,

Franklin,S., 1995, *Artificial Minds.* The Mit Press. Cambridge (Ma).

Gallay, A., 1989, Logicism: A French View Of Archaeological Theory Founded In Computational Perspective, *Antiquity* Vol. 63, 238, Pp. 27–39

Gardin, J.C., 1980, *Archaeological Constructs.* Cambridge: Cambridge University Press.

Gardin, J.C., 1991, *Le Calcul Et La Raison. Essais Sur La Formalisation Du Discours Savant.* Paris: Editions De L'ecole Des Hautes Etudes En Sciences Sociales.

Gardin, J.C., 1993, "Les Embarrass Du Naturel". *Archives Européennes De Sociologie Xxxiv, P. 152-165.*

Gardin, J.C., 1994, « Informatique Et Progrès Dans Les Sciences De L'homme » *Revue Informatique Et Statistique Dans Les Sciences Humaines.* 30 (1-4), Pp. 11-35.

Gardin, J.C., 1998, « Cognitive Issues And Problems Of Publication In Archaeology ». In *Theory And Practice Of Archaeological Research.* Edited By W.Hensel, S. Tabczynski, P. Urbanczyk. Institute Of Archaeology And Ethnology. Polish Academy Of Sciences. Warszawa.

Gardin, J.C., 2003, « Archaeological Discourse, Conceptual Modelling And Digitalisation : An Interim Report Of The Logicist Program". In *The Digital Heritage Of Archaeology.* Edited By M. Doerr And A. Sarris. Archive Of Monuments And Publications. Hellenic Ministry Of Culture.

Gardin, J.,C. Lagrange, M.S. Martin,, J.M., Molino,J., Natali-Smit,J., 1981, *La Logique Du Plausible. Essais D'épistemologie Pratique En Sciences Humaines.* Paris: Editions De La Maison Des Sciences De L'homme.

Gardin, J.C., Guillaume, O., Herman, P.O., Hesnard, A., Lagrange, M.S., Renaud, M., Zadora-Rio, E., 1987, *Systèmes Experts Et Sciences Humaines. Le Cas De L' Archéologie.* Paris: Eyrolles.

Gibbon, G., 1984, Kanthorpological Archaeology. New York, Columbia University Press.

Hall, J.S., 2007, Beyond Ai: Creating The Conscience Of The Machine, New York, Prometheus Books.

Hassanien, A.E., Abraham, A., Vasilakos, A.V., Pedrycz, W., (Eds.), 2009, ***Foundations Of Computational Intelligence*** 5 Vols. Berlin, Springer (Studies In Computational Intelligence)

Hendriks-Jansen,H.., 1996, *Catching Ourselves In The Act. Situated Activity, Interactive Emergence, Evolution, And Human Thought.* Cambridge (Ma), The Mit Press.

Holland, O. E., Mcfarland, D. 2001. *Artificial Ethology*. Oxford University Press

Hugget,J., Baker,K., 1986, "The Computerized Archaeologist: The Development Of Expert Systems" *Science And Archaeology*, 27: 3-12.

Humphreys, P.., 2004, *Extending Ourselves. Computational Science, Empiricism And Scientific Method.* New Yor, Oxford University Press.

Iida,F., Pfeiffer,R., Steels,L., Kuniyoshi,Y (Eds.) 2004, *Embodied Artificial Intelligence*. Berlin, Springer. Lecture Notes In Artificial Intelligence

Jones, T., 2007, *Artificial Intelligence: A Systems Approach* Hingham (MA), Infinity Press.

King,R.D., Whelan,K.E., Jones,F.M., Reiser,P.G.K., Bryant, C.H.K, Muggleton,S.H. Kell D.B., Oliver, S.G., 2004, Functional Genomic Hypothesis Generation And Experimentation By A Robot Scientist, *Nature* 427, 247 - 252

Kirsch,D., Maglio,P., 1995, "On Distinguishing Epistemic From Pragmatic Action". *Cognitive Science* 18, Pp. 513-549.

Klahr,D., 2000, *Exploring Science: The Cognition And Development Of Discovery Processes*, Cambridge (Ma), The Mit Press.

Klahr,D., Dunbar,K., 1988, Dual Space Search During Scientific Reasoning". *Cognitive Science* 12(1), Pp. 1-55.

Kovács, A.I., Ueno, H. 2005. The Case For Radical Epigenetic Robotics. *Proceedings Of The 10th International Symposium On Artificial Life And Robotics (Arob 2005)*, Beppu, Oita, Japan, 4.-6. February 2005. Available Online At Http://Www.Alexader-Kovacs.De/Kovacs05arob.Pdf

Kulkarni,D., Simon,H.A., 1988, "The Processes Of Scientific Discovery. The Strategy Of Experimentation". *Cognitive Science,* 12, 139-176.

Kurzweil, R., 2000, *The Age Of Spiritual Machines: When Computers Exceed Human Intelligence*. London, Penguin.

Lagrange,M.S., 1989, "Les Systèmes Experts Et La Récherche En Archéologie Et Sciences Humaines. Un Point De Vue Pragmatique". *Documentaliste* Vol. 26, N° 1, Pp. 11-15.

Langley, P., Simon, H.A., Bradshaw, G.L., Zytkov, J.M., 1987, *Scientific Discovery. Computational Explorations Of The Creative Process.* The Mit Press, Cambridge (Ma).

Liao, S.H., 2003, "Knowledge Management Technologies And Applications—Literature Review From 1995 To 2002". *Expert Systems With Applications* 25, Pp. 155–164

Liao,S.H., 2005, "Expert System Methodologies And Applications—A Decade Review From 1995 To 2004". *Expert Systems With Applications* 28, Pp. 93–103.

Luger, G.F., 2008, *Artificial Intelligence: Structures And Strategies For Complex Problem Solving* (6th Edition) Addison-Wesley.

Magnani, L., 2004, "Conjectures And Manipulations. Computational Modeling And The Extra- Theoretical Dimension Of Scientific Discovery". *Minds And Machines*, 14, Pp. 507–537.

Magnani, L., Dossena, R., 2006, *Computing, Philosophy And Cognition.* London, College Publications,

Marr,D.H., 1982, *Vision, A Computacional Investigation Into He Human Representation And Processing Of Visual Information* San Francisco, W.H. Freeman

Moravec,H., 20000, *Robot. Mere Machines To Trascend Mind*. Oxford University Press.

Munakata, T., 2008, *Fundamentals Of The New Artificial Intelligence: Neural, Evolutionary, Fuzzy And More*. Berlin, Springer (Texts In Computer Science).

Murphy, R.R., 2002, *Introduction To Ai Robotics*, Cambridge, The MIT Press.

Newell, A. 1982. The Knowledge Level. *Artificial Intelligence,* 18, 87-127.

Nolfi,S., Floreano,D., 2000, *Evolutionary Robotics. The Biology, Intelligence, And Technology Of Self-Organizing Machines*. Cambridge (Ma), The Mit Press.

Orlandi,T., 1997, "Informatica, Formalizzazione E Discipline Umanistiche". In *Discipline Umanistiche E Informatica. Il Problema Della Formalizzazione*, Edited By T. Orlandi, Roma, Pp. 7-17.

Orlandi, T., 2002, "Is Humanities Computing A Discipline?" *Jahrbuch Für Computerphilologie* 4, Pp. 51-58

Penrose, R., 1989, *The Emperor's New Mind: Concerning Computers, Minds And The Laws Of Physics*, New York, Oxford University Press.

Peschl, M.F., 1996, "The Development Of Scientific Concepts And Their Embodiment In The Representational Activities Of Cognitive Systems. Neural Representation Spaces, Theory Spaces, And Paradigmatic Shifts". *Philosophica* Vol. 57 (1), Pp. 131-172.

Pfeiffer,R., Scheier,C., 1999, *Understanding Intelligence*. Cambisge (Ma), The Mit Press.

Pollock, J.L., 1989, *How To Build A Person: A Prolegomenon* . Cambridge (Ma), The MIT Press.

Poole,D., Mckworth,A., Goebel,R., 1998, *Computational Intelligence. A Logical Approach.* Oxford, Oxford University Press.

Puyol-Gruart,J., 1999, "Computer Science, Artificial Intelligence And Archaeology". In *New Techniques For Old Times*. Edited By J.A. Barceló, I. Briz And A. Vila. Oxford, BAR Publishing (Bar International Series, 757, Pp. 19-27.

Read, D.W., 1985, The Substance Of Archaeological Analysis And The Mold Of Statistical Method: Enlightment Out Of Disdcordance. In Carr, C., (Ed.) 1985, *For* Concordance *In* Archaeological *Analysis: Bridging Data Structure Quantitative Technique And Theory*. Boulder, Westview Press. P. 45-86.

Read, D.W., 1990, The Utility Of Mathematical Constructs In Building Archaeological Theory. In Voorrips, A (Ed) *Mathematics And Information Science In Archaeology.* A Flexible Framework. P. 29-60. Holos Verlag (Studies In Modern Archaeology, Vol. 3), Bonn.

Russell, B., 1959, My Philosophical Development. London : Allen & Unwin,

Rutkowska, J.C., 1993, *The Computational Infant*. London, Harvester Wheatsheaf.

Santore, J.F., And Shapiro, S.C., 2004. A Cognitive Robotics Approach To Identifying Perceptually Indistinguishable Objects. In Alan Schultz, Ed., *The Intersection Of Cognitive Science And Robotics: From Interfaces To Intelligence, Papers From The 2004 Aaai Fall Symposium,* Technical Report Fs-04-05, Aaai Press, Menlo Park, Ca, 47-54.

Schiffer, M.B., 1976, *Behavioral Archaeology.* Orlando (Fl), Academic Press.

Segal, E., 1994, Archaeology And Cognitive Science In Colin Renfrew And Ezra Zubrow (Eds.) *The Ancient Mind: Elements Of Cognitive Archaeology* Cambridge University Press

Shennan,S.J., Stutt,A., 1989, "The Nature Of Archaeological Arguments". *Antiquity,* 64 (245), Pp. 766-777.

Shrager, J., Langley, P., (Eds.), 1990, *Computational Models Of Scientific Discovery And Theory Formation*. San Francisco (Ca), Morgan Kaufmann,

Simon,H.A., Lea,G., 1974, "Problem Solving And Rule Induction. A Unified View". In *Knowledge And Cognition*, Edited By L.W. Gregg. Lawrence Erlbaum Ass., Hillsdale (Nj).

Stary,C., Peschl,M.F., 1995, "Towards Constructivist Unification Of Machine Learning And Parallel Distributed Processing". In *Android Epistemology*. Edited By K.M. Ford, C. Glymour And P.J. Hayes. AAAI Press /The MIT Press. Menlo Park/Cambirdge/London.

Stein, L.A., 1995, "Imagination And Situated Cognition". In *Android Epistemology*. Edited By K.M. Ford, C. Glymour And P.J. Hayes. Aaai Press /The Mit Press. Menlo Park/ Cambirdge/London.

Stutt,A 1989, *Argument In The Humanities: A Knowledge-Based Approach.* Hcrl Technical Report No. 49. Milton Keynes: The Open University (Human Cognition Research Laboratory).

Tamburrini, G., Datteri, E., 2005, "Machine Experiments And Theoretical Modelling: From Cybernetic Methodology To Neuro-Robotics", *Minds And Machines* , Vol. 15, No. 3-4, Pp. 335-358.

Thagard, P., 1988, *Computational Philoophy Of Science*, Cambridge (MA), The MIT Press.

Trafton, J. G., Shultz, A. C., Perzanowski, D., Bugajska, M. D., Adams, W., Cassimatis, N. L., And Brock, D. P. (2004). Children And Robots Learning To Play Hide And Seek. *Cognitive Systems Journal*.

Tsaganou,G., Grigoriadou,M. Cavoura,T., Koutra,D., 2003, Evaluating An Intelligent Diagnosis System Of Historical Text Comprehension, Expert Systems With Applications 25, Pp. 493–502

Valdes-Perz, R.E., 1995, "Machine Discovery In Chemistry: New Results" *Artificial Intelligence,* 65 (2), Pp. 247-280.

Valdés-Pérez,R.E., 1996a, "Computer Science Research On Scientific Discovery" *Knowledge Engineering Review* 11, Pp. 57-66.

Valdés-Pérez,R.E., Zytkow,J., Simon,H.A., 1993, Scientific-Model Building As Search In Matrix Spaces" *Proceedings Aaai-93.* Washington,D.C.

Wagman, M., 2000, *Scientific Discovery Process In Humans And Computers*. Westport (Cn), Praeger Publishers

Wilcock, J., 1986, "A Review Of Expert Systems: Their Shortcomings And Possible Applications In Archaeology" *Computer Applications In Archaeology*

13, Pp. 139-144.

Wilcock,J., 1990, "A Critique Of Expert Systems, And Their Past And Present Use In Archaeology". In *Interpretation In The Humanities: Perspectives From Artificial Intelligence.* Edited By J.C. Gardin And R. Ennals. Library And Information Research Report, 71. The British Library Publications, Wetherby (Uk).

Winder,W., 1996, "Texperts". *Text Technology*, 6.3. Wright State University.

Winograd, T., Flores, F., 1985, *Understanding Computers And Cognition: A New Foundation For Design* Ablex Publ. Norwood Nj.

Zhang,C., Cao,C., Gu, F., Si, J., 2002, "A Domain-Specific Formal Ontology For Archaeological Knowledge Sharing And Reusing". In *Practical Aspects Of Knowledge Management* : 4th International Conference, Pakm 2002 Vienna, Austria. Proceedings. Edited By D. Karagiannis, U. Reimer. Springer-Verlag, Berlin. Lecture Notes In Computer Science, Vol. 2569, Pp. 213-225.

VIRTUAL IMPACT: Visualizing the Potential Effects of Cosmic Impact in Human History

W. Bruce Masse,[1] Maurizio Forte,[2] David R. Janecky[1] and Gustavo Barrientos[3]

[1] Los Alamos National Laboratory, USA

[2] UC Merced, USA

[3] Universidad de la Plata, Argentina

Abstract

Current models indicate that catastrophic impacts by asteroids and comets capable of killing more than one quarter of Earth's human population have occurred on average once every million years; smaller impacts, such the 1908 Tunguska impact that leveled more that than 2,000 square km of Siberian forest, occur every 200-300 years. Therefore, cosmic impact likely significantly affected hominine evolution and conceivably played a role in Holocene period human culture history. Regrettably, few archaeologists are trained to appreciate the nature and potential effects of cosmic impact. We have developed a conceptual model for an extensible set of educational and research tools based on virtual reality collaborative environments to engage archaeologists and the general public on the topic of the role of cosmic impact in human history. Our initial focus is on two documented asteroid impacts in Argentina during the period of 4000 to 1000 B.C. Campo del Cielo resulted in an energy release of around 2-3 megatons (100-150 times the Hiroshima atomic weapon), and left several craters and a strewn field covering 493 km^2 in northeastern Argentina. Rio Cuarto was likely more than 1000 megatons and may have devastated an area greater than 50,000 km^2 in central Argentina. We are focusing on reconstructions of these events and their potential effects on contemporary hunter and gatherers. Our virtual reality tools also introduce interactive variables (e.g., impactor physical properties, climate, vegetation, topography, and social complexity) to allow researchers and students to better investigate and evaluate the factors that significantly influence cosmic impact effects.

Key words: *virtual museum, asteroid impact, comet impact, natural disasters, ecosystem recovery, culture change*

1 Introduction

Disaster happens. And as evidenced by the enduring popularity of Hollywood disaster movies about earthquakes, exploding volcanoes, tsunami, burning sky scrapers, floods, hurricanes, sinking ships, tornados, nuclear war, alien invasions, overly rapid climate change, and even asteroid and comet strikes, many people clearly have an inherent fascination with the potential effects of natural and technological hazards (Kay and Rose 2006).

The publication in the *Proceedings of the National Academy of Sciences* on the hypothesized Younger Dryas impact around 12,900 BP has finally forced the archaeology community at large to take note of cosmic impact (Firestone et al. 2006; Kennett et al. 2009). If validated, the Younger Dryas impact may have been responsible for a dramatic 1300-year climate shift, megafaunal extinctions, and the rapid transformation of Clovis culture. Reactions in the archaeological community range between fascination, a strong desire to attack the possibility and to reject the evidence, and rare acceptance.

Planetary scientists who study near-Earth objects (NEO—asteroids and comets in orbits that threaten impact with the Earth), likewise express great skepticism toward the Younger Dryas and all other hypothesized recent major impacts (Kerr 2008; Pinter and Ishman 2008). As evidenced by the special session on the hypothesized Younger Dryas impact event at the 2008 Annual Meetings of the Society for American Archaeology, the recent cosmic impact debate is beset by discipline language barriers and by a lack of understanding on both sides of the nature of cosmic impact and the opportunities offered by its study (Masse 2007; Masse and Masse 2007; Masse, Weaver, Abbott, Gusiakov, and Bryant 2007; Barrientos and Masse 2009). Even more sobering is the realization that the hypothesized Younger Dryas impact could well be but one aspect of a much larger and more complex picture of the cosmic impact threat.

In in this paper, we explore ways in which cosmic impact can be better explained and visualized to the archaeological and anthropological community. The University of California Merced is developing a Virtual Heritage Center (VHC) based on virtual reality collaborative environments (display walls, virtual rooms and labs). A Virtual Museum (e.g., www.vhlab.itabc.cnr.it/flaminia) has been created inside the VHC, along with a simulation environment focused on Powerwall (display wall) technologies. The VHC provides scholars with new ways in which humanities-related data and knowledge can be searched, mined, displayed, taught, and analyzed. It employs multiuser domains (MUD), collaborative environments that allows a group of researchers or students to interact with each other and explore virtual worlds at the same time—unlike virtual reality systems that limit accessibility to a few users. Moreover, the VHC offers promise as a tool to teach students how to access, evaluate, and apply information in their studies as well as their lives beyond the academy.

But before plunging into this daunting challenge of figuring out how to apply this visualization technology to the study of cosmic impact, it is necessary to first make a few general observations about the relevance

Level I	Small disaster	< 10 persons	or	< 1 km^2
Level II	Medium disaster	10-100 persons	or	1-10 km^2
Level III	Large disaster	100-1000 persons	or	10-100 km^2
Level IV	Enormous disaster	1000-10,000 person	or	100-1000 km^2
Level V	Gargantuan disaster	> 10,000 persons	or	> 1000 km^2

Table 1. Disaster scope according to number of casualties and/or geographic area affected (modified from Gad-el-Hak 2008:Table 2.1).

and importance to archaeology of disaster research and cosmic impact studies.

2 Natural Disasters and the Significance of Cosmic Impact

We do not have to go back very far into historical records to establish disaster as a fact of life. For example, in the four decades between 1945 and 1986 more than 2.4 million people died as the result of disasters, an average of 30 disasters and 56,000 lives lost per year (cited in Torrence and Grattan 2002:1-2). While these numbers stagger the imagination, they pale beside the estimated 48 million people who died in the decade prior to 1945 as the direct or indirect result of war. Not even the 236,000 people who died in the December 2004 Indian Ocean earthquake and tsunami (Bryant 2008) can change the perception that human technological and social factors—such as war—have played a far larger role in causing death and culture change during the past 4000 years of recorded history than have the cumulative toll of all natural disasters.

However, this does not mean that the scientific study of natural hazards is of little consequence in our modern world. In fact, because natural disasters not only kill people but also significantly damage infrastructure resulting in many billions of dollars of economic loss, a sizable industry has developed to predict, control, and mitigate the risks of natural disasters and industrial accidents (e.g. Gad-el-Hak 2008). From this largely economic perspective, risk specialists have devised a heuristic logarithmic scale of the scope of a disaster, based on the number of casualties and/or the size of the geographic area affected (Gad-el-Hak 2008:Table 2.1). As depicted in Figure 1, the five tiers of such a scale range from a "small" disaster, resulting in the deaths of less that 10 individuals and affecting an area less than a square km, to that of a "gargantuan" disaster involving either the deaths of least 10,000 individuals or affecting an area greater than 1000 km^2.

Archaeologists and anthropologists increasingly have become involved with the study of natural disaster (e.g., Oliver-Smith and Hoffman 1999; Bawden and Reycraft 2000; Hoffman and Oliver-Smith 2001; Torrence and Grattan 2002; Grattan and Torrence 2007; Gould 2008). This topic was initially stimulated in part by the publication 30 years ago by Sheets and Grayson (1979) of their ground-breaking edited volume, *Volcanic Activity and Human Ecology*.

Ironically, however, the risk industry and archaeologists in particular, have generally ignored the largest by far of all natural hazards on Earth, that of the impact of asteroids and comets. Research on near-Earth objects (NEOs)—comets and asteroids in Earth-crossing orbits that pose the potential risk of impact on Earth—by default has been the purview of the small NEO community of planetary scientists. Most NEO community members think it unlikely that any humans have been killed by a cosmic impact in recorded history, and assume that because larger impacts are rare, impacts did not play a significant role in late Pleistocene and Holocene culture history. The following comment by the well-respected NEO specialist Clark Chapman (2008:418) is typical regarding historic cosmic impact:

> *"There have been reports of doubtful credibility from antiquity, as well as more recent anecdotes, of death by meteorite falls. While such an accident is certainly possible, there has been no confirmed, credible report of a human being dying from a meteorite strike."*

This remarkably simplistic and incorrect assumption is based on stochastic probabilities derived from the Solar System cratering record and the near-Earth asteroid population (e.g., Bottke 2007), and not on geological or archaeological and anthropological evidence, per se. In fact, such thinking is curiously antithetical to the NEO community's own models of cosmic impact risk. Globally catastrophic impacts capable of killing a quarter of Earth's human population (Chapman and Morrison 1994; Toon et al. 1997; Bobrowsky and Rickman 2007) are modeled to occur on average once per 500,000 to million years, well within the scope of human biological and cultural evolution. Such an impact generates an energy release $\geq 10^6$ megatons (MT), or roughly 67 million times that of the 15 kiloton Hiroshima and Nagasaki atomic weapons. At the other end of the scale are small impacts such as the 1908 Tunguska airburst which leveled 2150 km^2 of Siberian forest. These events occur on average every 100-200 years and yield 4-5 MT (4-5 x 10^{-1} MT), 267-334 times that of Nagasaki and Hiroshima (Boslough and Crawford 2008). In between are impacts with local (10^1-10^3 MT) and regional/continental (10^3-10^5 MT) effects, occurring on average between hundreds to hundreds of thousands of years.

General terrestrial impact effects (e.g., Melosh 2007) include thermal radiation as the projectile and target material are converted to incandescent gas or plasma;

seismic shaking at and away from the impact site; the deposition of ejecta from the impact; and the airburst created by displacement, compression, and heating of the air near the impact which produces ballistic shock waves. Effects can differ depending on the composition of the impactor (comets vs. iron-nickel, stony chondrite, and carbonaeceous asteroids); the pre-atmospheric speed (ca. 50 km/sec for comets, 17 km/sec for asteroids) and size of the impactor; the location of the detonation (airburst, land, ocean/lake, ice); the geology of the target site of a land impact (volcanics, granites, limestone, loess); and the angle of the impact (vertical to oblique).

In addition to the great complexity of all these variables, we have an especially poor understanding of airbursts and oceanic impacts. Oceanic impact results in the creation and collapse of a large water cavity (Figure 1). The subsequent infilling of the cavity actually tends to destroy or reshape aspects of the crater rim and surrounding ejecta sediment blanket due to the surging water column.

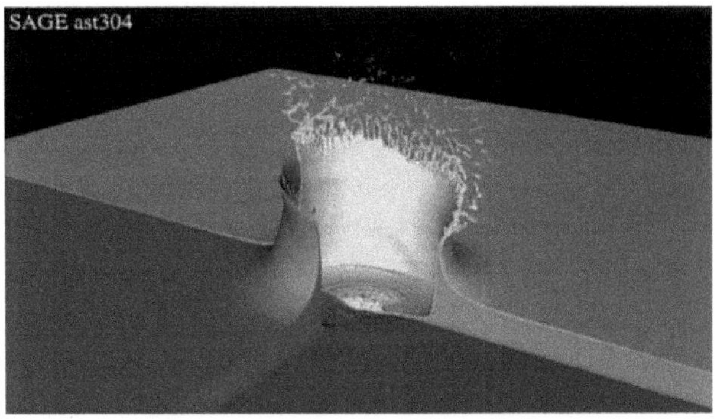

Figure 1. Three-dimensional simulation of the oceanic impact of a 1-km diameter asteroid.

IMPACTOR DIAMETER [METERS]	ENERGY RELEASE IN MEGATONS	CRATER DIAMETER (KM)	RECURRENCE INTERVAL (YEARS) BETWEEN IMPACTS	NUMBER OF WATER IMPACTS PER MILLION YEARS	NUMBER OF LAND IMPACTS PER MILLION YEARS	VALIDATED IMPACTS YOUNGER THAN A MILLION YEARS AS OF 2009* [IDENTIFIED THUS FAR ONLY ON LAND]
<36	<1	Airburst				(1) Haviland
36	1 1 MT is 67 times larger than Hiroshima	Airburst	132	5303	2273	(9) Sikhote Alin Wabar Sobolev Whitecourt Morasko Odessa Boxhole Dalgaranga Veevers
30	5	Airburst	200	3500	1500	Tunguska – Boslough and Crawford 2008
59	5	Airburst	413	1695	726	(4) Campo del Cielo Kaalijärv, Ilumetsä, Henbury
75	10^1	Airburst	718	975	418	(7) Barringer (Ariz.) Amguid Tswaing Kalkkop Wolfe Creek Monturaqui
210	10^2	1.6	7,700	91	39	(2) Tenoumer Lonar
360	10^3 [1 GIGATON]	4.3	27,000	26	11	(1) Rio Cuarto
730	10^4	8.3	140,000	5	2	
1550	10^5	16.3	780,000	1	0	(1) Zhamanshin
3330	10^6	32.1	4,600,000	0	0	
16 km	10^8	170	170 million	0	0	Chicxulub

*Validated Impacts derived from the University of New Brunswick Earth Impact Database as of April 2009 [http://www.unb.ca/passc/ImpactDatabase/].

Table 2. Hypothesized impact rates during the past one million years for stony (chondritic) asteroids with an impact velocity of 17 km/second and an impact angle of 45 degrees. [Adapted from the Earth Impact Effects Program, Collins et al. 2005, with the exception of **Tunguska** estimate].

Table 2 uses the University of Arizona Earth Impact Effects Program (Collins et al. 2005) to calculate a hypothetical logrithmic distribution of impacts during the past one million years. For ease of comparison, the calculated impacts are based upon the general parameters of composition as a stony (chronditic) asteroid entering Earth's atmosphere at a velocity of 17 km/second and an angle of trajectory of 45 degrees.

This table illustrates that many cosmic impacts should have occurred during the past million years, but more importantly that the vast majority have not yet been identified in the geological and archaeological record (as indicated by the small numbers of validated impacts in the last column). The potential impacts include many larger impacts that should have left substantial evidence in the geologic record. Table 2 also demonstrates that the issue of identifying oceanic impacts is particularly thorny in that no impact structure from any magnitude of impact (including globally catastrophic impacts) has yet been identified on the sea floor from the past 25 million years, although a few tektite strewnfields have been recorded, likely indicative of a nearby impact.

3 The Campo del Cielo and Rio Cuarto Argentina Asteroid Impacts

The Campo del Cielo (Field of the Sky) crater field is in northeastern Argentina (Figure 2). This portion of the Gran Chaco is semi-arid, hot, very flat, and covered equally by savanna, scrub, and dense thorn forests. The Campo del Cielo crater field contains at least 20 small, generally elongated impact craters within a northeast-trending ellipse (N63°E), 4 km wide and 19.2 km long (Cassidy et al. 1965; Cassidy and Renard 1996; Liberman et al. 2002; Wright et al. 2007).

The largest crater is 115 x 91 m. The largest surviving meteorite, dubbed El Chaco, is more than 2 m in diameter and weighs an estimated 37,000 kg (Figure 3). The shallow impact in-fall angle was calculated at 9° from the horizontal. The main concentration of craters is at the southwestern end of the ellipse. The four largest and least elongated craters were formed by ground explosions spread along a 6-km line. The smaller craters did not experience explosive impact but instead contain meteorite penetration funnels, four of which have been excavated.

Calculated impact velocities range from 1.7–4.3 km s^{-1}. Analysis of impact effects yielded a pre-atmospheric entry velocity of 22.8 km s^{-1}, a diameter greater than 6.0 m, and mass minimally at 840,000 kg. Impact glass melts have not been recovered from the Campo del Cielo crater field, likely indicative of the small size and relatively slow speed of the impactors. An elliptical strewn field extends 60 km beyond the craters, and small meteorites have been recovered all along the path.

Three charcoal specimens were recovered from Campo del Cielo (Cassidy and Renard 1996: Table 4). Two samples bracket the impact, while the other is believed to be from

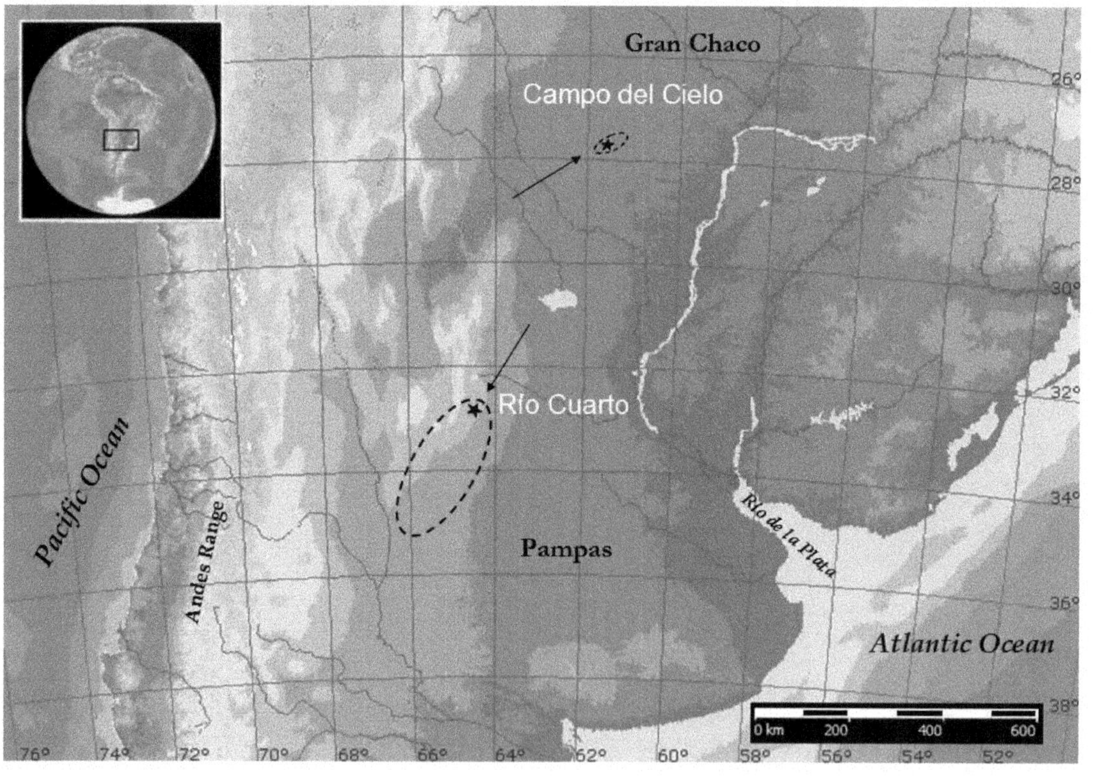

Figure 2. Location of the Campo del Cielo and Rio Cuarto crater fields in Argentina. The stars represent the crater fields, the arrows the estimated direction of impact infall, with the dashed line for Rio Cuarto representing the approximate limits of the melt glass strewnfield.

Figure 3. The Campo del Cielo "El Chaco" meteorite.

an ignition fire caused by the impact itself. The most recent sample was recovered from Crater 1 post-impact infill, yields a calibrated (2σ) date range of A.D. 900-1431 (Barrientos and Masse 2009). The earliest sample, from a soil horizon buried by the ejecta blanket of Crater 2, yields a calibrated (2σ) date range of 5209-4264 B.C. The third sample was recovered from the beginning of the impact tunnel at the bottom of Crater 10 and yields a calibrated (2σ) date range of 2840-2146 B.C. That this sample may date the actual impact is consistent with the location of Crater 10 near the middle of the distribution of explosive craters (Wright et al. 2007:Fig. 1).

Figure 4. The Rio Cuarto crater field illustrated in 1992 on the cover of the journal *Nature*.

Río Cuarto, in central Argentina (Figure 4), has been proposed as a Holocene shallow angle asteroid impact zone containing at least 11 elongated craters (Schultz and Lianza 1992). The largest is about 1.1 x 4.5 km. The region contains extensive deposits of loess and dune sand which can be altered by wind and readily melted by cosmic impact. Schultz and Lianza (1992) noted that the Rio Cuarto structures could be interpreted to be of aeolian origin; however, they also note the presence of impact glass melts (locally referred to as *escorias* or slag-like rocks) in and around some of the putative craters (Figure 5).

A series of alternate interpretations have been proposed in response to the original claims of impact formation for the Río Cuarto elongated depressions. Cione et al. (2002) argue for a non-impact origin, and state that Argentine geologists have long supported the aeolian formation of these depressions, pointing out that the alignments match prevailing wind patterns during the period(s) of their formation. Bloom (1992) has argued that the glass melts are unlikely to be of impact origin. He suggests that natural and anthropogenic fires, such as the burning of fields as part of the agricultural cycle, can produce the melts described by Schultz and Lianza (1992).

Using aerial photography, Bland et al. (2002) also suggest that the Río Cuarto structures are part of a widespread set of several hundred elongated aeolian depressions associated with parabolic sand dunes that formed in the Argentine Pampas during the mid-Holocene. Unlike Bloom (1992) and Cione et al. (2002), Bland et al. (2002) support the impact origin of the Río Cuarto glasses; however, their interpretation is that the glasses represented the distal ejecta of a Pleistocene (ca. 480 ka B.P.) age impact occurring several hundred kilometers away.

Figure 5. Holocene *escoria* glass melt from Rio Cuarto. The white bar scale is 1 cm.

Schultz and his colleagues (2004) are part of an ongoing systematic NSF-funded study to understand the nature and distribution of the impact glasses found in the Argentine loess. Their research includes detailed petrological and electron microprobe analyses, along with geochemical analyses (x-ray florescence and instrumental neutron activation analysis) of the Argentine glass melts. They demonstrated that temperatures in excess of 1700ºC would be necessary to completely melt all constituents including quartz grains. The specimens exhibit rapid quenching of the melt, a condition unlikely in a field fire or wildland fire. Also, the glasses are in stratigraphically-restricted contexts, unlike the expected more widespread and haphazard distribution if they were instead the result of lightning or field fires.

Masse and Masse (2007) note that the physics of wildland fire—and by analogy field fire—also argues against the anthropogenic burning of fields as the origin of the glass melts as suggested by Bloom (1992) and San Cristóbal (1999). Pyne et al. (1996:20-23) state that the theoretical maximum temperatures that can be achieved by the burning of combustible gases generated from wildland fuels is around 1900–2200 °C. However, most wildland fires more typically burn at average temperatures of between 700–980 °C with the maximum actually measured for an exceptionally intense fire being not much greater than 1650 °C, below the melt threshold determined by Schultz et al. (2004) for the Río Cuarto glass melts. In addition, the burning of fields in preparation for agriculture, including the presence of smoldering fires, would typically yield temperatures lower than the average wildland fire due to the differences in fuels.

In their initial exhaustive analysis of the melt glasses from the Pampas, Schultz et al. (2004) identified melts from five separate Quaternary impacts with four of the impacts dating back to the Pleistocene period between 570 ± 100 ka B.P. (corresponding to the material identified by Bland and his colleagues) and 114 ± 26 ka B.P., along with a Holocene glass melt. Schultz and his colleagues (2006) also identified two separate late Miocene vesicular impact glasses from the Pampas, for a current total of eight distinct late Cenozoic impact melt breccia deposits dating between 9.24 Ma and 6 ka B.P. (Harris and Schultz 2007).

Of significance is the robust documentation of the age of the Holocene impact glasses by three dating techniques: 10,000-4000 B.P. based on geological context (stratigraphy and preservation state); 2300 ± 1600 B.P. by fission track dating; and 6000 ± 2000 B.P. by radiometric $^{40}Ar/^{39}Ar$ dating. The combined suite of dating techniques implies a date for the impact of approximately 6000-3000 B.P., encompassing the date range for Campo del Cielo.

The Campo del Cielo has a crater field of approximately 45 km^2, and a meteorite strewn field of perhaps another 448 km^2 (assuming an ellipse 9.5 x 60 km). Comparisons with other known and intensively studied small impacts, such as Tunguska, the 1947 Sikhote Alin meteorite shower also in Siberia, and the 49,000 year old Meteor Crater in Arizona suggests that the impact energy release of Campo del Cielo was likely around 2-3 MT (Barrientos and Masse 2009). Because of the considerable size of the craters and strewn fields, the atmospheric airburst detonations would have been several kilometers above the ground surface It is uncertain if a fireball was created by the atmospheric detonation (...mythology suggests that a fireball and associated debris cloud did occur, as noted below), but impact ground temperatures were apparently insufficient to create glass melts. We suggest an average fatality rate of 50% for any humans who may have been present within the boundaries of the 493 km^2 Campo del Cielo impact crater and strewn field (presumably a higher percentage in the crater field, lower in the meteorite strewn field).

The Río Cuarto impact event is both qualitatively different from and quantitatively larger than Campo del Cielo, and indeed likely several magnitudes larger than the Tunguska event. The way we presently understand the modeling and interpretations of P. Schultz and his colleagues and those of their critics (see above), there are four different rough scenarios by which to interpret Río Cuarto (Barrientos and Masse 2009), with the accumulated data strongly supporting the following two scenarios.

The first scenario, consistently favored by P. Schultz over the years (pers. comm. 2008) based in part on extensive modeling and simulation associated with oblique projectiles (e.g., Schultz 1991), accepts the reality of the original hypothesized Río Cuarto crater field and views the distribution of Holocene glass melts as most likely representing impactites spread out by the explosive contact that created the craters. If the relationship between the crater field location and orientation and the defined Holocene glass melt distribution (Figure 2) continues to be supported by the geologic data—and Schultz (pers. comm. 2008) wisely cautions that there is still much work to be done to verify this apparent relationship—then this would strongly suggest that the craters and melts are indeed intimately related, and that an area of nearly 50,000 km^2 was seared by the impact, much like a rolling nuclear explosion. The low angle of entry may have also caused devastation on the ground by the bow shock wave in front of and below the impactor for many kilometers prior to actual impact.

The second scenario views the Holocene glass melts as the in situ residue of an airburst, rather than as ejecta from a cratering event. Airbursts are objects that explode and release most of their energy in the atmosphere above the Earth's surface, such as the Tunguska event. The frequency and magnitude of airbursts and other classes of cosmic impact are still critical topics of research (Bland and Artemieva 2003; Morrison et al. 2003; papers in Bobrowsky and Rickman 2007), as demonstrated by the recent downsizing of the magnitude of Tunguska and increasing the frequency of this class of impact event. Wasson (2003) suggests that airbursts can be created by weakly-structured objects much larger and therefore much more energetic than that of Tunguska, exceeding 1000 MT. An example of such an object would be Comet Shoemaker-Levy 9 that due to tidal forces from Jupiter broke up into 21 separate fragments before colliding with the planet in July 1994. Wasson argues that a super-Tunguska impact could produce such a large fireball on detonation that the ground surface over a brief period of time would reach temperatures greater than 2000 °C, thus melting anhydrous soils such as loess and dune sand. Boslough and Crawford (2008) have similarly modeled chondritic asteroids capable of producing similar effects but with considerably less energy necessary to produce the airburst glass melts. Wasson (2003) suggests that his largest modeled airbursts could impact surface areas greater than 100,000 km^2, a figure twice that of the tentatively defined Río Cuarto Holocene glass melt field.

It is difficult for us to choose between the first scenario (Río Cuarto cratering impact) and the second scenario (airburst), but both scenarios better fit our existing data than do scenarios that deny the reality of the impact or that the glass melts are the result of crater formation elsewhere in central Argentina. The airburst and cratering impact scenarios seemingly imply more than 1000 MT of energy release (based on current modeling programs such as that by Collins et al. 2005). Not only would the Rio Cuarto devastation be considerably larger than the area devastated by the Campo del Cielo impact, but also the mortality rate in these areas should be much greater than that calculated for the Campo del Cielo impact, perhaps approaching 90% fatalities. That this conservative figure is not set at 100% takes into account topography and other situational and physical variations of the impact event.

Using data published by Binford (2001), some approximations about relevant parameters of the kind of societies likely involved in the Rio Cuarto impact can be made. From his database of 339 hunter-gatherer groups, Barrientos and Masse

(2009) isolated the 19 that inhabit the area between 32º and 36º of latitude (both north and south) in order to calculate the approximate number of individuals and cultural groups or societies which could be directly affected by the impact. The hunter-gatherer societies that lived at that latitudinal strip during historical times (6 at different parts of Australia, and 13 in the American Southwest) exhibited a quite restricted residential mobility (0-9 annual movements, with a median of 0.1 and a interquartile range between 0 and 7), a diet mainly composed by gathered resources (with a median of 50%), and a highly variable total population size (a median of 2,124 people, with a interquartile range between 889 and 3,500 people) and population density (a median of 18.0 inhabitants/100 km^2, with a interquartile range between 5.12 and 43.75 inhabitants/100 km^2).

The hunter-gatherer societies living in central Argentina at the time of the impact event are believed to have been mostly low-dense at the regional scale of analysis. Taking this into account, some filtering of the data was necessary in order to get more realistic approximations (Barrientos and Perez 2005). First, after standardizing the quantitative data, a *k*-means cluster analysis was performed using three of the variables coded by Binford (2001:117)—total population, population density per 100 km^2, and the average number of annual residential moves so as to analytically separate two more internally homogeneous groups. The analysis discriminated between a first cluster of 8 cultural groups with higher total population, higher population density and lower residential mobility, and a second cluster composed by 11 groups with lower total population, lower population density, and higher residential mobility. The members of the latter cluster were selected for further analysis.

Considering that the area directly affected by the impact was approximately an ellipse of 48,106 km^2, a median number of 3 (interquartile range between 1 and 11) cultural groups or societies with such characteristics may have been extirpated by this sudden, unpredictable event. This implies a median number of about 1,000 individuals (interquartile range between 500 and 1,908) affected in some way by the impact, 900 of which (90%) probably died. Applying this demographic model to Campo del Cielo, and our estimate of 50% lethality for the impact area results in a death toll of around 5 individuals. These estimates could be even much higher, especially for Campe del Cielo, because they do not include the effects of fast-moving wildland fires radiating out from the impact zones, which would have killed additional individuals and further depleted the resources necessary for human maintenance and recovery.

Summarizing, the two mid-Holocene cosmic impact events in Argentina had the potential of being very destructive. However, the Río Cuarto impact would seem to be both qualitatively different from and quantitatively larger than Campo del Cielo as well as the most likely potential contributor to any major environmental and population disruption in central Argentina during the mid-Holocene. Using the criterion of size of the affected areas as suggested in the disaster scope scale outlined in Table 1, the Campo del Cielo impact would be categorized as an "enormous" disaster while that of Rio Cuarto would be categorized as a "gargantuan" disaster.

4 Using Mythology to Investigate the Possibility that the Campo del Cielo and Rio Cuarto Impacts Created Major Wildland Fires

Planetary scientists have debated the capacity of impacts to start ignition fires (e.g., Jones and Lim 2000; Svetsov 2002; Jones 2002; Durda and Kring 2004). Witnessed meteorite falls, and the limited archaeological record of impact sites support the notion that wildland fires are associated with some smaller impacts. A key would be fuel availability and suitable weather/climatic conditions for the spread of the fire. Mythology also provides a surprising window on the relationship between Campo del Cielo and Rio Cuarto and potentially devastating wildland fires.

During the past few decades, there has been a growing trend to use the natural sciences to re-examine the context, structure, and meaning of myth (e.g., Vitaliano

1973; Blong 1982; Barber and Barber 2004; Piccardi and Masse 2007; Masse, Barber, Piccardi, and Baber 2007; Cashman and Giordano 2008; Van der Sluijs 2009). This approach demonstrates that there is a powerful historical core to myth that can be objectively retrieved and studied. The messages and data contained in myth are not just simple curiosities but instead provide important insights into the development of human cognition in relation to both science and religion.

Even more critical is the fact that catastrophe myths contain a record of significant natural disasters (earthquakes, tsunami, volcanic eruptions, floods, plagues, cosmic impacts) in recent Earth history. These can be used to provide insights for natural hazard analysis and planning, and can even be applied to the question of whether or not the Earth is significantly at risk from impact by comets and asteroids. Myth data suggest that cosmic impact has played a significant role in recent cultural evolution.

Valuable for our interests in cosmic impact are a set of 4,259 myths from 20 major cultural groups east of the Andes gathered by the University of California at Los Angeles (Wilbert and Simoneau 1992). These myths are the contribution of 111 authors, translated into English as necessary, and published over the course of 20 years as a set of 23 separate volumes. The cultural groups themselves are from widely distributed portions of South America (Figure 6).

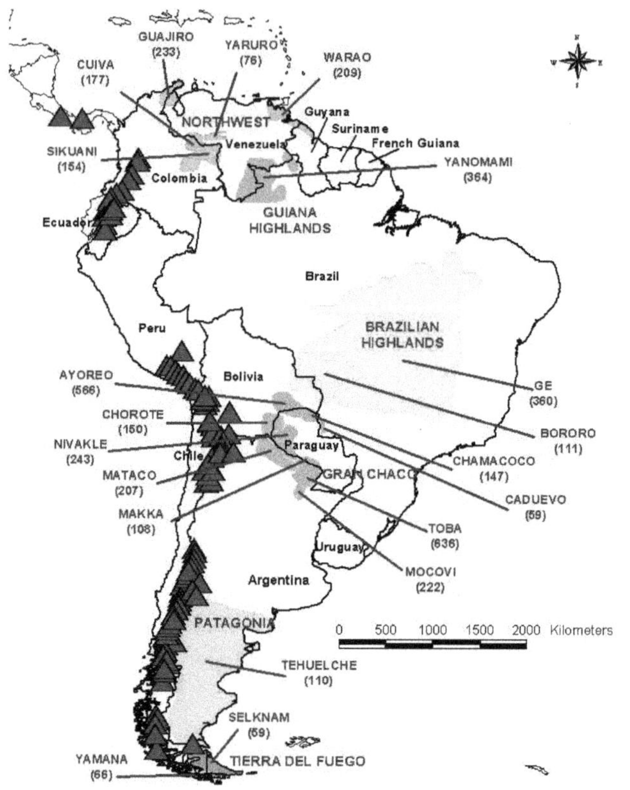

Figure 6. Distribution of cultures and associated myths in the UCLA *Folk Literure of South America* series.

Masse and Masse (2007) analyzed these 4,259 myths, concentrating on those describing various local, regional, or worldwide natural catastrophes that led to the deaths of members of a given cultural group. Events that led to the deaths of small numbers of individuals include local floods, fire, lightning—and in the case of two myths from the Brazilian Highlands, the observed thunderous fall of a meteorite into a river that killed several youths then swimming in the river. Several other myths from the Brazilian Highlands, and the Northwest talk about meteorites as being capable of causing human death—including, poignantly, the overall eventual destruction of the world—but do not describe actual impact events themselves.

Of particular interest is a set of 284 myths that have as their primary motif a single major cataclysm stated as having led to the deaths of most or all members of one or more cultural groups—typically referred to as having led to new creations of humanity (Table 3). While one might scoff at the rational basis of such "new creations," it should be remembered that these cultural groups typically were small, a few hundred or at most a couple thousand people, and that while their overall territorial ranges may have been large the cultural group only occupied a small portion at any given time. Therefore, rare large-scale cataclysms such as plinian eruptions, mass fires, and torrential monsoons of unusual duration could indeed decimate such groups.

Table 3 organizes these myths by cultural group and by five defined categories of cataclysm. The stories, particularly those from the Gran Chaco, appear to be divided into relative time within the overall set of 284 myths, with certain cataclysms being stated as having occurred before or after other cataclysms. Thus myths about a lengthy time of darkness and a similar set of myths combining darkness with the sky falling or collapsing on top of people, houses, and forests are said to have occurred most recently (but still in the distant past), while myths of a "great" or "worldwide fire" occur in the middle of the myth cycle, and myths about a "great flood" occur at the beginning of the myth cycle, the latter sometimes coupled with a period of "great cold" stated as having occurred immediately after the flood.

Elsewhere it has been shown that the recent "sky fall" and "great darkness" myths likely capture a major Plinian eruption of Bolivia's Nuevo Mundo volcano during the past two thousand years prior to the arrival of Europeans in South America (Masse and Masse 2007). The myths of a great flood and great cold arguably relate to the observed effects of an oceanic comet impact around 4800 years ago (Masse 1998, 2007, 2009).

The myths of the world fire are of interest here for four basic reasons. First, although these stories are almost entirely limited to the South American regions most subject to natural wildland fire and anthropogenic field burning (i.e., the Gran Chaco and the Brazilian Highlands), the myths describe the unique rapid spread and severity

CULTURE & LOCATION	GREAT FLOOD [EARLIEST IN MYTH CYCLE]	GREAT COLD [AFTER FLOOD MYTH]	WORLD FIRE [MIDDLE OF MYTH CYCLE]	SKY FALL–DARKNESS [LATEST IN MYTH CYCLE]	GREAT DARKNESS [LATEST IN MYTH CYCLE]
NORTHWEST					
Cuiva	13			4	
Guajiro	9				1
Sikuani	10				
Warao	3				
Yaruro	10				
GUIANA HIGHLANDS					
Yanomani	17			11	2
BRAZILIAN HIGHLANDS					
Bororo	4		1		
Ge	11		6		1
GRAN CHACO					
Ayoreo	17		2		1
Caduevo					
Chamacoco	10		1	1	
Chorote	10		7	1	
Makka	2				1
Mataco	12		5		1
Mocovi	7		3		
Nikvale	6	1	6	9	
Toba	24	5	27		12
PATAGONIA					
Tehuelche	2				
TIERRA DEL FUEGO					
Selknam	1				
Yamana	3	1	2		1
TOTALS	171	7	60	26	20

Table 3. "Worldwide" catastrophe myths in the UCLA *Folk Literure of South America* series.

of the fire(s). The following two Gran Chaco myths, respetively from the the Toba and Mataco groups are particularly illustrative:

> The people were all sound asleep. It was midnight when an Indian noticed that the moon was taking on a reddish hue. He awoke the others: "The moon is about to be eaten by an animal" [a lunar eclipse]. The animals preying on the moon were jaguars, but these jaguars were spirits of the dead. The people shouted and yelled. They beat their wooden mortars like drums, they thrashed their dogs....They were making as much noise as they could to scare the jaguars and force them to let go their prey. Fragments of the moon fell down upon the earth and started a big fire. From these fragments the entire earth caught on fire. The fire was so large that the people could not escape. Men and women ran to the lagoons covered with bulrushes. Those who were late were overtaken by the fire. The water was boiling, but not where the bulrushes grew. Those who were in places not covered with bulrushes died and there most of the people were burnt alive. After everything had been destroyed the fire stopped. Decayed corpses of children floated upon the water. A big wind and a rain storm broke out. The dead were changed into birds. The large birds came out from corpses of adults, and small ones from the bodies of children [Métraux 1946:33; in Wilbert and Simoneau 1982a:68].

> Once, a very long time ago, the life-style of the Mataco developed into near anarchy... All was chaos. Then one day a big black cloud gathered in the south, and lightning and thunder began. When the cloud had covered the entire sky it began to rain a bit here and there, but the drops that fell were not like rain but like fire. The people tried to jump into the river to save themselves, but the water was boiling. Tokhuah [a Mataco trickster character] was among them, but he saved himself because he could go wherever he wished, and he decided to go underground. All died but a very few, and they did not know why they had survived. Bits of fire continued to fall from the sky and everything, including the entire forest, burned; nothing remained except a few people here and a few there. They did not understand why they were still alive and what they were to do next [Fock, in Wilbert and Simoneau 1982b, p. 126].

The second reason that these stories are of interest is that the cultures with these stories are hundreds of kilometers removed from the nearest active Holocene volcanoes (the red triangles in Figure 6), thus largely eliminating explosive eruption as a potential natural cause for the storyline descriptions. In this regard it is noted that migration stories are largely absent from the oral traditions of these cultures, thus suggesting that the Gran Chaco and Brazilian Highlands are the likely setting for the storied events.

Third, the Campo del Cielo crater-strewn field is less than 100 km south of the edge of the historic distribution of the Toba and Mataco tribes.

The fourth and most obvious reason is that the Toba provide a very specic cause for the events described in the world fire myths:

> Moon...is a pot-bellied man whose bluish intestines can be seen through his skin. His enemy is a spirit of death, the celestial Jaguar. Now and then the Jaguar springs up to devour him. Moon defends himself with a spear tipped with a head carved of the soft wood of the bottletree..., which breaks apart at the first impact. He also has a club made of the same wood which is too light to cause any harm. The Jaguar tears at his body, pieces of which fall on the earth. These are the meteors, which three times have caused a world fire [Métraux 1946:19].

Analysis of the complete set of world fire myths (Masse and Masse 2007; Barrientos and Masse 2009) indicate a clear link between these myths and the Campo del Cielo impacts, and possibly also that of Rio Cuarto and a hypothesized airburst in the Brazilian Highlands.

5 Capturing Cosmic Impact in a Virtual Museum

Based on the above discussion, it is readily apparent that topics such as natural disasters, cosmic impact, and natural science-based mythologies are extraordinarily complex and inherently imbued with language and methods of study that are difficult for the physical sciences, natural sciences, and social sciences to bridge and meaningfully discuss. These, then, are the challenges that must be faced and surmounted in attempting to present such topics in a virtual reality visualization format.

Our proposed project explores a common educational theme in studies of virtual reality—that of finding visually creative ways in which to interest people in a body of scientific (in this case archaeological) data. However, rather than simply trying to interest a group of students or the general public in appreciating and understanding archaeological sites and landscapes as historically interesting features, we have chosen a case study that presents unusual real world challenges in terms of bridging differences between cultures and crossing barriers between scientific disciplines. Not only does this attempt at bridging create technological challenges as to how virtual reality data are viewed and processed, but the research questions that lie at the heart of this case study have profound implications for the study of cultural and biological evolution.

We envision that our methodological approach for the visualization of cosmic impact is to create a network of virtual heritage and collaborative environments between UC Merced, Los Alamos National Laboratory (LANL), the University of La Plata, and other cooperating institutions, reconstructing the past through simulation virtual processes and multiuser digital spaces.

Our scenario is to have collaborative environments for students, scholars and visitors where the users can interact in a 3D virtual space exchanging data and information of eco-cultural and archaeological content. Advanced research, education, communication, learning will be the key words representing the contents of the virtual heritage. The term "virtual heritage" means the virtual ontology of the cultural heritage. In practical terms, the VHC model should open new important perspectives for a network of excellence and for a matrix of research in this field.

The ultimate outcomes of this collaborative research project are summarized in the following six key actions:

- Experimenting with advanced simulation processing within virtual reality collaborative environments between different labs and research teams.
- Teaching and focusing research through virtual collaborative environments.
- Reconstructing sites and event sequences of the past, stressing the relationships between cultures, ancient societies, environmental data and ecosystems.
- Creating pilot case studies for approaching new methodological guidelines in the field of virtual heritage and in the validation of the reconstruction and dynamic simulation process of the past, within the framework of an ecological perspective.
- Integrating skills, methods and competences of different labs and cross-disciplinary teams.
- Simulating by virtual immersive environments complex phenomena related with archaeological and environmental datasets.

6 The Steps for Capturing Cosmic Impact in a Virtual Museum

Our case study builds on research that examines the effects of the Campo del Cielo and Río Cuarto asteroid impacts on 3rd millennium B.C. Argentine hunting and gathering societies (Masse 1998, 2007; Masse and Masse 2007; Barrientos and Masse 2009). Except for limited work on the late Holocene Kaali impact in Estonia (Veski et al. 2007), and the 800,000 bp Australasian tektite strewfield impact on *Homo erectus* (Langbroek and Roebroeks 2000; Langbroek 2004), and recent precocious studies in relation to the hypothesized impact at the beginning of the Younger Dryas climatic ordeal 12,900 years ago (Firestone et al. 2007; Kennett et al. 2009), there has been no systematic study of a validated cosmic impact by archaeologists. The biggest hurdle will be that of demonstrating to archaeologists that the methods and theories of the archaeological discipline can actually contribute to a better understanding of cosmic impact.

We envision the following steps towards the establishment of a working "virtual impact" environment.

Step I. Gathering detailed information and the creation of a virtual reality package involving core cosmic impact textual concepts and associated visual illustrations is a necessary basis, augmented where possible with animations and pertinent internet links. Among these would be: (1) Discussion of the Solar System including the nature and orbital characteristics of comets and asteroids. (2) Pertinent definitions such as meteor, meteorite, bolide, meteor showers and storms, etc. (3) Presentation of the visually-impressive known cratering record in our Solar System outside of the Earth itself. (4) Discussion of the nature, effects, and implications of the 65.5 million year old Tertiary-Cretacious (KT) geologic period boundary Chicxulub impact. (5) A review of the major types of impact events/processes known on Earth, including airbursts, simple terrestrial impact structures, complex terrestrial impact structures, and oceanic impacts. (6) A review of the known cratering record on Earth. (7) A discussion of the process of validating impact structures. (8)
A review of the current accepted astrophysical model of impact rates on Earth. (9) Visualizations highlighting the rather profound discrepancy that exists between the models of impact rates and our current validated lists of impacts.

Thorough integration of these aspects of astrophysics and interactions with the Earth will provide an intellectually stimulating environment for scientific exploration and learning.

Step II. The study of the Rio Cuarto and Campo del Cielo sites and their potential to inform on impact processes is enhanced by local mythology and oral tradition that has captured important aspects of the cosmic impacts, such as the likelihood that the impacts created largescale wildland fires. Thus a discussion and set of visualizations of a natural science approach to mythology and oral tradition, geomythology, and related topics will be integrated into our project.

Step III. Simulation of the Mid-Holocene Argentina impact events would be built, including physical effects and processes. This includes a discussion of the overall Argentine impact record based on the work of Peter Schultz and his colleagues (e.g., Schultz et al. 2004, 2006). The Campo del Cielo case study builds from the modeling and visualization already accomplished by William Cassidy (University of Pittsburgh) and his colleagues. The Rio Cuarto case study will build upon the modeling and visualization already accomplished by Peter Schultz (Brown University) and his colleagues.

We anticipate that colleagues at Los Alamos National Laboratory (LANL) may be able to contribute to our virtual impact visualizations of the Rio Cuarto and Campo del Cielo impacts by building on previously designed "SAGE" codes, developed by LANL and Science Applications International Corporation (SAIC) that have simulated aspects of cosmic impact in both 2D and 3D format.

A majority of large simulations come in one of two constructs: Lagrange, in which a grid or mesh of mathematical points matches with and follows molecules or other physical variables through space; or Eulerian, in which the mesh is fixed in space, thereby permitting researchers to follow fluids as they move from point to point.

The power of the SAGE (SAIC's Adaptive Grid Eulerian) code lies in its flexibility. The mesh can be continuously refined to increase the level of detail the code provides about specific physical elements in the mesh. Newer SAGE simulations use increasingly realistic equations to represent physical variables such as the atmosphere, seawater and ocean crust when modeling, for example, tsunami propagation from oceanic impact, such as illustrated in Figure 1.

Although a goal of such simulation is that of developing better and better resolution of event physical processes, application of the data from these huge calculations by broad multi-disciplinary sets of researcher require excellent visualization and interactive work techniques, such as the three-dimensional power walls used at LANL and UC Merced.

These SAGE models will then be applied to the reconstructed environment and human demography of the mid-Holocene Campo del Cielo and Rio Cuarto impacts. The preliminary modeling of Barrientos and Masse (2009) will be expanded to better capture the potential effects of these two impacts on their contemporary human societies.

With sufficient data gathered from the Argentine paleoenvironment and human demography, it will also be possible to introduce simulations structured on agent-based models. For example, researchers at the Santa Fe Institute during the past decade have developed relatively robust tools such as the "Artificial Anasazi Project" (Gumerman et al. 2002). This simulative experiment was designed to provide empirical evaluation of the principles and procedures associated with a bottom-up, agent-based computer simulation to illuminate human behavior in a real world setting.

A potential Artificial Argentina Impact Project could create landscapes from reconstructed environmental variables and would be populated by families, households, or tribal bands as artificial agents. It would draw upon detailed environmental reconstructions of the Rio Cuarto and Campo del Cielo impact sites as impinged upon by low- and high-frequency paleoenvironmental variability in order to study how changing physical and social conditions play out in terms of long-term settlement patterns. We then could systematically alter the quantitative parameters or make qualitative changes that introduce subtle, completely new, and even unlikely elements into the artificial world of the simulation.

In this specific case we would introduce the effects of asteroid impact vis-à-vis Campo del Cielo or Rio Cuarto. By allowing the alteration some of the physical variables of impact, we can use visulation technology to explore and better understand those parameters most likely to affect regional ecosystems and the demographic characteristics of imbedded human populations.

Step IV. This is a long-term visionary step that would attempt to build on present largely static interactive programs that help to visualize the physical effects of impact in order to achieve dynamic interactive programs focusing on variables pertaining to ecosystems and human population dynamics.

For example, a current popular and useful web-based program for estimating the regional environmental consequences of an impact on Earth is that of the Earth Impact Effects Program (Marcus, Melosh, and Collins 2005), which was used to estimate the impact rates listed in Table 2. The Earth Impact Effects Program estimates the ejecta distribution, atmospheric blast wave, ground shaking, and thermal effects of an impact as well as the size of the crater producted. This is determined by choosing a stipulated distance from the impact, and then inputting a specific series of impactor parameter such as the size and density of the impactor projectile, the impact velocity, the degree of impact angle, and the nature of the target location such as water depth if an oceanic or lake impact, sedimentary rock (e.g. loess, sand, sandstone), or crystalline rock such as surface volcanic magma flows and metamorphic rocks.

It would be useful to enhance the visualization of and interaction with information derived from the Earth Impact Effects Program by the use of 3D GIS topography layers. For example, the program would benefit by the use of Google Earth-like imagery that translate the calculated effects into visualized landscapes based on a range of specific target locations.

More critical is the fact that the Earth Impact Effects Program does not include most of the significant physical effects associated with oceanic impact, a significant omission in light of the fact that water and ice cover 70% of the Earth's surface. These effects not only include the obvious potential for tsunami propogation, but also include such things as the amount and types of particulates injected into the upper atmosphere.

Impacts larger than 1000 megatons may produce striking environmental and climatic effects in addition to those physical effects described above (Toon et al. 1997; Birks et al. 2007; MacCraken 2007). Particularly poorly understood are deepwater oceanic impactors in the range of 10^4-10^7 megatons in terms of their injection of large quantities of particulates and water vapor into the upper atmosphere that may trigger and sustain long-lived cyclonic storms, including super-hurricanes.

In a similar vein, since ocean surface heating is a major factor in initiating El Niño climatic events, could a modest impact (10^3-10^5 megatons) in a sensitive oceanic location trigger a major El Niño event? And what would happen if a larger-sized impact hit the glacial ice of Greenland or Antarctica? Not only would this inject a large amount of water vapor so as to considerably disturb the atmosphere, but how might the pulse of melted freshwater effect ocean currents, atmospheric-oceanic couplings, and oceanic food webs?

We also have much to learn about the potential signature(s) of impact in the global climate change record, including tree-rings, ice cores, and paleobotanical and pollen records (e.g., Baillie 1994, 2007).

Also missing from these physical parameters for the Earth Impacts Effects Program are parameters that deal with the potential for igniting or spreading wildland fires such as vegetation type (tropical forest, open woodland, savanna, taiga); climatic conditions (e.g., drought, monsoon, temperate); terrain/topography categorization; fuel loads (such as used in wildland fire matrices); temperature; time of day; and wind conditions.

And critically missing are ecosystem and human demographic and social complexity variables that are key for calculating effects of impacts on human populations. What types of societies are present in the impact target area? Are they hunter and gatherers, semi-sedentary or sedentary agriculturalists, urban civilizations, or a mix of these gross societal categories? What is the population density, and are populations clustered or evenly scattered out over the landscape?

Catastrophic impacts constitute a largely unstudied bottlenecking mechanism from a human and biological evolutionary perspective. Just exactly what does "globally catastrophic" or any other magnitude of cosmic impact mean for past cultures and ecosystems? We need to focus on the role of cosmic mega-killers ($\geq 10^6$ MT) in human evolution (e.g., Australasian and Younger Dryas impacts); on possible extinction thresholds for smaller impacts on hunting and gathering societies (e.g., Rio Cuarto); on the ways in which impacts of various sizes might have caused the collapse of past complex societies (e.g., Courty et al. 2008); and on the potential evolutionary role of recent cosmic impact in climate and ecosystem change. Archaeologists and kindred historical and evolutionary disciplines can also contribute to dialog regarding issues of disaster planning and recovery, which typically omit consideration of cosmic impact (e.g., Gad-el-Hak 2008).

7 Conclusions

Archaeologists conduct fieldwork worldwide and employ exquisite micro-stratigraphic, paleoenvironmental, and chronological controls to explore major aspects of human history. We model population dynamics and cultural behavior. And we deal with documentary records and oral traditions. This places us in an ideal position to scrutinize and evaluate the Pleistocene and Holocene historical and sedimentary record of cosmic impact, and to fill in some of the gaps that currently exist in the planetary science community's models of the rates, risks, and effects of impact.

This new research endeavor likely will at first seem very foreign and intimidating to many archaeologists and for most members of our kindred discipline of anthropology. We do tend to be very nervous when it comes to working with data and methods requiring application of detailed knowledge of physics, mathematics, and chemistry.

The Virtual Impact visualization package is intended to break down some of the apparent barriers that now currently exist on the road to this new line of historical inquiry. We suggest that not only will it serve as a useful tool as archaeologists begin to consider how cosmic impacts have affected past human populations and ecosystems, but also it can open doors for additional fruitful dialog with the physical and planetary science communities.

Acknowledgements

The authors thank Peter Schultz for graciously sharing his knowledge and interpretations concerning the Rio Cuarto impact. Robert Weaver and Galen Gisler provided useful discussions of their 2D and 3D SAGE code models of oceanic impact. Los Alamos National Laboratory has assigned Technical Information Release No. LA-UR-09-xxxxx to this paper.

Bibliography

Barber, Elizabeth Wayland, and Paul T. Barber. *When They*

Severed Earth from Sky: How the Human Mind Shapes Myth. Princeton: Princeton University Press, 2004.

Barrientos, Gustavo, and Sergio I. Perez. "Was There a Population Replacement During the Late Mid-Holocene in the Southeastern Pampas of Argentina? Archaeological Evidence and Paleoecological Basis." *Quaternary International* 132 (2005):95-105.

Barrientos, Gustavo, and W. Bruce Masse. "Mid-Holocene Asteroid Impacts in Central Argentina: Exploring Probable Effects on Human Population Dynamics." Revised paper sumbitted to *American Antiquity*, November 2009.

Bawden. Garth, and Richard Martin Reycroft (Editors). *Environmental Disaster and the Archaeology of Human Response*. Albuquerque: Maxwell Museum of Anthropology, Anthropological Papers No. 7, 2000.

Binford, Lewis R. "*Constructing Frames of Reference: An Analytical Method for Archaeological Theory Building Using Ethnographic and Environmental Data Sets*. Berkeley: University of California Press, 2001.

Birks, John W., Paul J. Crutzen, and Raymond G. Roble. "Frequent ozone depletion resulting from impacts of asteroids and comets." In: Bobrowsky, P.T., and H. Rickman (Editors) *Comet/Asteroid Impacts and Human Society: An Interdisciplinary Approach*, pp 225-245. Berlin: Springer, 2007.

Bland, P. A., and N. A. Artemieva. "Efficient disruption of small asteroids by Earth's atmosphere." *Nature* 424 (2003):288-291.

Bland, P. A., C. R. Souza Filho, A. J. T. Jull, S. P. Kelley, R. M. Hough, N. A. Artemieva, E. Pierazzo, J. Coniglio, L. Pinotti, V. Evers, and T. Kearsley. "A Possible Tektite Strewn Field in the Argentinian Pampa." *Science* 296 (2002):1109-1111.

Blong, Russell J. *The Time of Darkness: Local Legends and Volcanic Reality in Papua New Guinea*. Seattle, University of Washington Press, 1982.

Bloom, A. " A non-impact explanation for elongated depressions near Río Cuarto, Córdoba Province, Argentina." In *Geological Society of America Meeting, Abstracts and Programs*, A136-A137, 1992.

Bobrowksy, Peter, and Hans Rickman (Editors). "*Comet/ Asteroid Impacts and Human Society*." Berlin: Springer, 2007.

Boslough, M.B.E., and D.A. Crawford. "Low-Altitude Airbursts and the Impact Threat." *International Journal of Impact Engineering* 35 (2008):1441-1448.

Bottke, William F., Jr. "Understanding the Near-Earth Object Population: The 2004 Perspective." In *Comet/ Asteroid Impacts and Human Society*, edited by Peter Bobrowsky and Hans Rickman, pp. 175-187. Berlin: Springer, 2007.

Bryant, Edward A. *Tsunami: The Underrated Hazard.* Chichester: Springer/Praxis Publishing, 2008.

Cashman, K.V., and G. Giordano (Editors). "Volcanoes and Human History." Special Issue of the *Journal of Volcanology and Geothermal Research* 176 (2008):325-438.

Cassidy, William A., and Marc L. Renard. "Discovering Research Value in the Campo del Cielo, Argentina, Meteorite Craters." *Meteoritics & Planetary Science* 31 (1996):433-448.

Cassidy, William A., Luisa M. Villar, Theodore E. Bunch, Truman P. Kohman, and Daniel J. Milton. "Meteorites and Craters of Campo del Cielo, Argentina. *Science* 149 (1965):1055-1064.

Chapman, Clark R. "Meteoroids, Meteors, and the Near-Earth Object Impact Hazard." *Earth, Moon, and Planets* 102 (2008):417-424.

Chapman, Clark R., and David Morrison. "Impacts on the Earth by Asteroids and Comets: Assessing the Hazard." *Nature* 367 (19940:33-40.

Cione, Alberto L., Eduardo P. Tonni, Jorge San Cristóbal, Pedro J. Hernández, Adrián Benítez, Fernando Bordignon, and Jorge A. Perí. "Putative Meteoritic Craters in Río Cuarto (Central Argentina) Interpreted as Eolian Structures." *Earth, Moon and Planets* 91 (2002):9-24.

Collins, Garreth S., H. Jay Melosh, Robert A. Marcus. "Earth Impact Effects Program: A Web-based Computer Program for Calculating the Regional Consequences of a Meteoroid Impact on Earth." *Meteoritics & Planetary Science* 40 (2005):817-840.

Courty, M.-A., A. Crisci, M. Fedoroff, K. Grice, P. Greenwood, M. Mermoux, D.Smith, and M. Thiemens. "Regional manifestation of the widespread disruption of soil-landscapes by the 4 kyr BP impact-linked dust event using pedo-sedimentary micro-fabrics." In *New Trends in Soil Micromorphology,* edited by S. Kapur, A. Mermut, and G. Stoops, pp. 211-236. Berlin: Springer, 2008.

Durda, Daniel D., and David A. Kring. "Ignition threshold for impact-generated fires." *Journal of Geophysical Research* 109 (2004):1-14.

Firestone, R. B., A. West, J. P. Kennett, L. Becker, T. E. Bunch, Z. S. Revay, P. H. Schultz, T. Belgya, D. J. Kennett, J. M. Erlandson, O. J. Dickenson, A. C. Goodyear, R. S. Harris, G. A. Howard, J. B. Kloosterman, P. Lechler, P. A. Mayewski, J. Montgomery, R. Poreda, T. Darrah, S. S. Que Hee, A. R. Smith, A. Stich, W. Topping, J. H. Wittke, and W. S. Wolbach. "Evidence for an Extraterrestrial Impact 12,900 Years Ago that Contributed to the Megafaunal Extinctions and the Younger Dryas Cooling." *Proceedings of the National Academy of Sciences* 104 (2007):16016–16021.

Gad-el-Hak, Mohamed. "The art and science of large-scale disasters." In *Large-Scale Disasters: Prediction, Control, and Mitigation*, edited by Mohamed Gad-el-Hak, pp. 5-68. Cambridge University Press, Cambridge, 2008.

Grattan, John, and Robin Torrence (Editors). *Living Under the Shadow: The Cultural Impacts of Volcanic Eruptions*. Walnut Creek: Left Coast Press, 2007.

Harris, R. Scott, and Peter H. Schultz. "The Record of Late Cenozoic Impacts in the Argentine Pampas: Consequences of Hypervelocity Collisions into Soft Sedimentary Target. Paper 137-4. *Geological Society of America Abstracts with Programs*, vol. 39, No. 6

(2007):371.

Hoffman, Susanna, and Anthony Oliver-Smith (Editors). *Catastrophe and Culture: The Anthropology of Disaster*. Santa Fe: School of American Research Advanced Seminar Series, 2002.

Jones, Timothy P. "Reply 'Extraterrestrial impacts and wildfires'". *Palaeogeography, Palaeoclimatology, Palaeoecology* 185 (2002):407-408.

Jones, Timothy P., and Bo Lim. "Extraterrestrial Impacts and Wildfires". *Palaeogeography, Palaeoclimatology, Palaeoecology* 164 (2000):57-66.

Kay, Glenn, and Michael Rose: *Disaster Movies*. Chicago: Chicago Review Press, 2006.

Kennett, D.J., J.P. Kennett, A. West, C. Mercer, S.S. Que Hee, L. Bement, T.E. Bunch, M. Sellers, and W.S. Wolbach. "Nanodiamonds in the Younger Dryas Boundary Sediment Layer". *Science* 323 (2009):94.

Kerr, Richard A. "Experts Find No Evidence for a Mammoth-Killer Impact. *Science* 319 (2008):1331-1332.

Krinov, Yevgeny L. *Giant Meteorites*. Oxford: Pergamon Press, 1966.

Langbroek, M. "Narrowing Down the Significance of a Large Cosmic Impact Catastrophy in Southeast Asia for Early Asian *Homo erectus*." In Langbroek, M. *'Out of Africa,' an Investigation into the Earliest Occupation of the Old World.* BAR International Series 1244, pp. 104-108. Oxford: BAR Publishing, 2004.

Langbroek, M., and W. Roebroeks. "Extraterrestrial Evidence on the Age of the Hominids from Java." *Journal of Human Evolution* 38 (2000):595-600.

Liberman, R. G., F. Niello, M. L. di Tada, L. K. Fifield, J. Masarik, and R. C. Reedy. "Campo del Cielo Iron Meteorite: Sample Shielding and Meteoroid's Preatmospheric Size. *Meteoritics Planetary Science* 37 (2002):295-300.

Masse, W. Bruce. "The Celestial Basis of Civilization." *Vistas in Astronomy* 39 (1995):463-477.

Masse, W. Bruce. "Earth, Air, Fire, & Water: The Archaeology of Bronze Age Cosmic Catastrophes." In *Natural Catastrophes During Bronze Age Civilisations: Archaeological, Geological, Astronomical, and Cultural Perspectives*, edited by B. J. Peiser, T. Palmer and M. E. Bailey, pp. 53-92. BAR International Series 728. BAR Publishing, Oxford, 1998.

Masse, W. Bruce. "The Archaeology and Anthropology of Quaternary Period Cosmic Impact." In *Comet/Asteroid Impacts and Human Society*, edited by Peter Bobrowsky and Hans Rickman, pp. 25-70. Springer Press, Berlin, 2007.

Masse, W. Bruce. "Using Myth to Hypothesize and Define the World's Greatest Historic Catastrophe: The Flood Comet Impact." *Technique and Technology*, Siberian Federal University, 2009, (in press).

Masse, W. Bruce, and Michael J. Masse. "Myth and Catastrophic Reality: Using Myths to Identify Cosmic Impacts and Massive Plinian Eruptions in Holocene South America." In *Myth and Geology*, edited by L. Piccardi and W. B. Masse, pp. 177-202. London: Geological Society of London Special Publication 273, 2007.

Masse, W.B, R.P. Weaver, D.H. Abbott, V.K. Gusiakov, and E.A. Bryant. "Missing in Action? Evaluating the Putative Absence of Impacts by Large Asteroids and Comets During the Quaternary Period. *Proceedings of the 2007 Advanced Maui Optical and Space Surveillance Technologies Conference*, pp. 701-710, Wailea, Maui, Hawaii, 2007.

Masse, W. Bruce, Elizabeth Wayland Barber, Luigi Piccardi, and Paul T. Barber. "Exploring the Nature of Myth and its Role in Science. In *Myth and Geology*, edited by L. Piccardi and W. B. Masse, pp. 9-28. Geological Society of London Special Publication 273, London, 2007.

Melosh, H. Jay. "Physical Effects of Comet and Asteroid Impacts: Beyond the Crater Rim." In *Comet/Asteroid Impacts and Human Society*, edited by Peter Bobrowsky and Hans Rickman, pp. 211-224. Berlin: Springer, 2007.

Morrison, David, Alan W. Harris, Geoff Sommer, Clark R. Chapman, and Andrea Carusi. "Dealing with the impact hazard." In *Asteroids III*, edited by William Bottke, Alberto Cellino, Paolo Paolicchi, and Richard P. Binzel, PP.739-754. Tucson: University of Arizona Press, 2003.

Oliver-Smith, Anthrony, and Sussana M. Hoffman (Editors). *The Angry Earth: Disaster in Anthropological Perspective*. New York: Routledge, 1999.

Piccardi, Luigi, and W. Bruce Masse (Editors). "*Myth and Geology*." London: Geological Society of London Special Publication 273, 2007.

Pinter, Nicholas, and Scott E. Ishman. "Impacts, Mega-Tsunami and Other Extraordinary Claims. *GSA Today* 18 (2008):37-38.

Pyne, Stephen J., Patricia L. Andrews, and Richard D. Laven. *Introduction to Wildland Fire*, 2nd edition. New Tork: John Wiley & Sons, Inc., 1996.

San Cristóbal, Jorge O. "El más probable origen de las escorias y tierras cocidas." In *Resúmenes de las XV Jornadas Argentinas de Paleontología de Vertebrados*, pp. 24, 1999.

Schultz, Peter H. "Atmospheric effects of oblique impacts." *Abstracts of Papers Submitted to the Lunar and Planetary Science Conference* 22 (1991):1191-1192.

Schultz, Peter H., and Robert E. Lianza. "Recent Grazing Impacts on the Earth Recorded in the Rio Cuarto Crater Field, Argentina. *Nature* 355 (1992):232-237.

Schultz, Peter H., Marcelo Zárate, Bill Hames, Christian Koeberl, Theodore Bunch, Dieter Storzer, Paul Renne, and James Wittke. "The Quaternary Impact Record from the Pampas, Argentina. *Earth and Planetary Sciences Letters* 219 (2004):221-238.

Schultz, Peter H., Marcelo Zárate, Willis E. Hames, R. Scott Harris, T.E. Bunch, Christian Koeberl, Paul Renne, and James Wittke. "The Record of Miocene Impacts in the Argentine Pampas." *Meteoritics & Planetary Science* 41 (2006): 749-771.

Sheets, Payson, and Donald Grayson. *Volcanic Activity and Human Ecology*. New York: Academic Press, 1979.

Svetsov, Vladimir V. "Comment on 'Extraterrestrial impacts and wildfires'." *Paleogeography, Paleoclimatology, Paleoecology* 185 (2002):403-405.

Toon, Owen B., Kevin Zahnle, David Morrison, Richard P. Turco, and Curt Covey. "Environmental Perturbations Caused by the Impacts of Asteroids and Comets." *Reviews of Geophysics* 35 (1997):41-78.

Torrence, Robin, and John Grattan. *Natural Disasters and Cultural Change*. London: Routledge, 2002.

Van der Sluijs, Marinus Anthony. "Book Review of Piccardi, L. and W.B. Masse (eds.), 2007. *Myth and Geology.* Geological Society Special Publication 273. London: The Geological Society." *Myth and Symbol* 5 (2009):58-74.

Veski, Siim, Atko Heinslau, Anneli Poska, LeiliSaarse, and Jüri Vassiljev. "The Physical and and Social Effects of the Kaali Meteorite Impact—A Review. In *Comet/ Asteroid Impacts and Human Society*, edited by Peter Bobrowsky and Hans Rickman, pp. 265-275. Berlin: Springer, 2007.

Vitaliano, Dorothy B. *Legends of the Earth: Their Geologic Origins.* Bloomington: University of Indiana Press, 1973.

Wasson, John T. "Large Aerial Bursts: An Important Class of Terrestrial Accretionary Events." *Astrobiology* 3 (2003):163-179.

Wilbert, Johannes, and Karin Simoneau (Editors). *Folk Literature of the Toba Indians* (Vol. 1). Los Angels: UCLA Latin American Center Publications, University of California, 1982a.

Wilbert, Johannes, and Karin Simoneau (Editors). *Folk Literature of the Mataco Indians*. Los Angeles: UCLA Latin American Center Publications, University of California, 1982b.

Wilbert, Johannes, and Karin Simoneau (Editors). *Folk Literature of South American Indians: General Index.* Los Angeles: UCLA Latin America Center Publications, University of California, 1992.

Wright, S.P., M.A. Vesconi, M.G. Spagnuolo, C. Cerutti, R.W. Jacab, and W.A. Cassidy. "Explosion Craters and Penetration Funnels in the Campo del Cielo, Argentina Crater Field. *Lunar and Planetary Science* 38 (2007), No. 2017.

PLACE-Hampi, Ancient Hampi and Hampi-LIVE - an entanglement of people-things

Sarah Kenderdine

Special Projects, Museum Victoria, Melbourne/ Visiting Associate Professor, City University, Hong Kong

Abstract

This paper describes two immersive interactive installations (PLACE-Hampi and Hampi-LIVE) and one exhibition (Ancient Hampi: the Hindu Kingdom Brought to Life). These works have sought to redefine interactive narrative for the interpretation of tangible and intangible aspects of the UNESCO World Heritage site of the Monuments of Hampi, in South India. The discussion of these projects (completed between 2006 and 2010) draws upon theories in post-processural archaeology (Interpretive Archaeology and Symmetrical Archaeology) to describe the potential for new media tools to contribute to reframing of engagement with cultural heritage for multi-cultural audiences. This paper also acknowledges the potential for archaeology as 'theatre' to describe the performative qualities of immersive installations such as PLACE-Hampi and Hampi-LIVE as a tricotomy of relationships between users (performers), the system and, spectators. This research is supported by extensive qualitative and quantitative evaluation of PLACE-Hampi and will be summarized briefly here.

Key words: Interpretive Archaeology, Symmetrical Archaeology, Phenomenology, Interactive Technologies, Hampi, India, Evaluation

1 Introduction[1]

The interactive installations PLACE-Hampi and Hampi-LIVE contribute to a reframing of cultural heritage interpretation and the reformulation of somatic, kinaesthetic[2] and embodied experiences in immersive digital environments. The research has been articulated through transdisciplinary experiments in two distinct large-scale, interactive cinematic architectures (hereafter 'immersive architectures'). These immersive architectures and their associated visual, sonic and algorithmic techniques offer compelling means for mapping and remediating the tangible, intangible and abstract aspects of culture and heritage landscapes. The installations are an inquiry in multisensory journeying that addresses perceived challenges in visualizing 'things' in archaeology, people in landscapes and the imaginary of texts. The works act as propositions for the expressions of digital narrative and aesthetics through virtual embodiment and enaction.[3] This research both contributes to and exploits emerging technologies.

Throughout the arts and sciences, new media technologies are allowing practitioners the opportunity for cultural innovation and knowledge transformation. These practitioners are straining at their disciplinary boundaries to provide products for the 'information society', 'knowledge economy' and 'creative industries'. As media archaeologist Siegfried Zelinski incites, the challenge for contemporary practitioners engaged with these technologies is *not* to produce "more cleverly packaged information of what we know already, what fills us with ennui, or what merely serves laziness and tries to harmonize what is not yet harmonious."[4] Instead, Zelinski celebrates those who, inside the laboratories of current media praxis, understand the invitation to experiment and to "continue working on the impossibility of the perfect interface."[5] To research the "perfect interface" is to work across heterogeneity and to encourage "dramaturgies of difference."[6]

The research practice involving new forms of immersive architectures described in this paper[7] is motivated by both Zelinski's call for "dramaturgies of difference" and the potential for the remediation of archaeological and historical landscapes and culture knowledge. The theoretical framework that underpins these investigations is partly drawn from the discourses of post-processual archaeology, including the domains of Interpretive and

[1] The title 'entanglement of people-things' is inspired by Christopher Witmore's discussion on symmetry in archaeology (cf Bruno Latour) and Timothy Webmoor's T. 2007, 'Lessons From the Real: Mediating People-things in a Symmetrical Archaeology' (2007), available online <http://traumwerk.stanford.edu:3455/Symmetry/743> last accessed 30 June 2009. See also Witmore, C. L. 'Four archaeological engagements with place: Mediating bodily experience through peripatetic video', in *Visual Anthropology Review*, vol. 20, no. 2, (2004): 57–72.

[2] A body in motion is one endowed with kinaesthetics, derived from the Greek *kinein*, meaning 'to move' and *aesthesia*, 'to perceive'.

[3] Action and perception are intertwined in complex ways. Perception controls action; action helps perception. A current trend in cognitive studies endows action with a wide role in determining perception. The term enaction has been coined to characterize this peculiar approach to perception. In the context of virtual heritage see also Forte, M. & Bonini, E., 'Embodiment and enaction: A theoretical overview for cybercommunities', Proceedings *of Virtual Systems and Multimedia* 08, Cyprus (2008).

[4] Zelinski, S. *Deep time of the media: Toward an archaeology of hearing and seeing by technical means*, Custance, G. (trans.), (Cambridge, MA: MIT Press 2006), 280.

[5] Ibid. 259.

[6] Ibid.

[7] The author is involved in research and development for a number of immersive architectures including The Virtual Room <http://www.vroom.org.au>, iDome <http://icinema.unsw.edu.au/projects/infra_dome.html>, Advanced Visualisation Environment <http://icinema.unsw.edu.au/projects/infra_avie.html>, Re-Actor <http://unmakeablelove.org/>, The EYE, <http://www.vrmag.org/vartist/spotlight/THE_EYE_OF_NAGAUR.html> and PLACE <http://place-hampi.museum>.

Symmetrical Archaeology. From fissures within the long-established practices of processural archaeology, there has been a growing call to "contest traditional predominantly textual modes of thought and action" and to "illuminate and expose the interpretive and *artistic* qualities of presentation and narration" using visual media.[8] Similarly, advocates of media [in] archaeology[9] suggest broader approaches to interpretation that move toward an archaeological expression of "recombinatory poetics."[10]

In the post-processual frameworks for Interpretive Archaeology, practitioner Christopher Tilley advances a phenomenological understanding of the experience of landscape.[11] His arguments emphasize approaches to archaeological assemblages that go beyond their material manifestation. In his book Body and Image: Explorations in Landscape Phenomenology, Tilley usefully contrasts iconographic approaches to the study of representation with those of kinaesthetic enquiry. Tilley's line of reasoning provides grounding for the research into large-scale, immersive and kinaesthetically provocative environments, designed for the experience and embodiment of cultural and heritage. In this context, it is useful to quote his discussion of the experiential interpretation of rock art: "The potential of kinaesthetic approaches tell us something different... Iconographic approaches are usually primarily cognitive in nature. They grant the primacy to the human mind as a producer of the meaning of the images through sensory perception. It is the mind that responds in a disembodied way... Kinaesthetic approaches, by contrast, stress the role of the carnal human body. The general claim is that the manner in which we perceive, and therefore relate to visual imagery, is fundamentally related to the kinds of bodies we have. The body both limits and constrains and enables us to perceive and react to imagery in specific embodied ways."[12]

Such arguments form crucial foundations for the works discussed in this essay. The challenge for digital cultural heritage disciplines is to design environments where, for example, "archaeologists take up what is left of the past and translate and (re)mediate."[13] As Shanks encourages these new strategies "avoid the predefinition of data itself, evidenced in the current meta-endeavour, and moves towards an approach that destabilizes the notion of finding what we are looking for."[14]

The immersive installations that contribute to this essay follow a methodological framework that incorporates an understanding of the body in motion, in both real and digital spaces. By suggesting that landscape studies are transformed by re-focusing on the kinaesthetic, Tilley stresses the autonomy of images to convey meaning. Imagery has a "direct agency"[15] that acts as one inhabits and *moves* through the landscape. The design of PLACE-Hampi and HAMPI-Live demands that people ambulate and circumambulate, continuously re-orienting themselves in relation to real-world scale imagery of augmented virtual landscapes. The acoustic spaces in these systems are dynamic in relation to the positioning of visitors and their movements. Participants interact and perform with *both* the imagery of the virtual world and with the other people who co-inhabit the space. The issues of inhabiting the immersive space and its relationship to real environments and the performative qualities of people within these hybrid and multimodal spaces are part of the discussion in this essay.

New forms of archaeological investigation draw upon, and advance, performance theory and practice.[16] Archaeology can be considered in "terms of *different relationships* with what is left of the past."[17] This foregrounds the anthropological questions of the performance and construction of the past in memory and of narrative meaning. These processes involve a revisioning of collections (texts and material objects) and the redesign of archives and systems of documentation. The two installations described below can be interpreted partly as theatres for new modalities of performance in archaeological contexts.

The investigative frameworks described by the work in this paper address a complex mix of human-computer interface issues that arise during the hermeneutic,[18] phenomenological and epistemological[19] encounters with archaeological and cultural heritage landscapes. Such methodology recognizes that the past is constantly re-created in learning experiences, a never-ending process of 'making sense'. This supports the notion that interpretation is always in flux is fundamental

[8] Cochrane, A. & Russell, I. 2007, 'Visualizing archaeologies: A manifesto', in *Cambridge Archaeological Journal*, vol. 17, no. 1 (2007): 3–19.

9 For example, see Webmoor, T., 'Mediational techniques and conceptual frameworks in archaeology', in *Journal of Social Archaeology*, vol. 5, no. 1, (2005): 54–86. Shanks, M. 2006, *Media and archaeological futures*, available online <http://documents.stanford.edu/michaelshanks/113>, last accessed 19 May 2009.

[10] In this paper recombinatory poetics is '...the combining of materials in the digital realm, from different sources and of different orders, in such ways that they resonate to create meanings within the spaces of their combination as much as in the elements themselves' (Shanks 2006 *op cit*).

[11] Refer specifically to Tilley, C. *The phenomenology of landscape*, (Oxford: Berg 1994). Tilley, C. *The materiality of stone: Explorations in landscape phenomenology*, (Oxford: Berg, 2004). Tilley, C. *Body and image: Explorations in landscape phenomenology,* (Walnut Creek: Left Coast Press, 2008).

[12] Tilley, C. 2008, op cit, 18.

[13] Shanks, M. 2008, *Archaeology and the visual*, available online <http://documents.stanford.edu/michaelshanks/144>, last accessed 30 June 2009.

[14] Ibid.

[15] Tilley 2008, op cit. 46.

[16] Pearson, M. & Shanks, M. *Theatre/Archaeology*, (London: Routledge 2001). Also, *The Presence Project*, available online <http://presence.stanford.edu:3455/Collaboratory/19>, last accessed 30 June 2009.

[17] *The Presence Project* op cit.

[18] See in particular the hermeneutic spiral, which is defined as a living tradition of re-creation. It is fundamentally a matter of perceiving a moving horizon, engaging a strand of dialogue that is an on-going re-articulation of the dynamically historical nature of all human thought. See Hodder 1999 and Thomas 2004.

[19] See renowned anthropologist and cyberneticist Gregory Bateson for his articulation of ecological epistemology (Bateson 1972, 1979, Bateson & Bateson 1988).

for two reasons: firstly, it reinforces the experiential real-time encounter with interactive systems, wherein no participant will experience the same thing twice, even after recurring visits to the system. Secondly, it strengthens the personalized cognitive and kinaesthetic encounter as a valid form of interpretation—which is supported by narrative agency.

In the research described narrative content and immersive architectures combine to provide a context for multisensory mediation between humans, machine agents and virtual environments. Modalities of interaction can be characterized as forms of prosthetic vision, acoustic immersion, kinaesthetic activation, telepresence, inhabitation and dwelling, travelling, driving and walking, and dynamic contemplation. In the cultural imaginary activated by immersive architectures, the sensory world of participants is tuned for encounter and thus emergent narratives become possible.

The research into immersive architectures for cultural heritage also investigates relationships between embodiment and representations of the material. Don Ihde, post-phenomenologist and philosopher of science and technology, promotes a *material* hermeneutics that 'gives things voices where there had been silence, and brings to sight that which was invisible'.[20] The 'things' of Idhe's visual hermeneutics are viewed through instrumental magnification of prosthetics, such as telescopes and microscopes, thus allowing perception to go where it has not gone before. Likewise, the immersive architectures described in this essay promote the visualization of the material—of 'things' *and* aspects of embedded meanings emanating from those things. The prosthetic devices within the interactive display amplify the sensorial and allow the senses to go where they have not gone before. Museum specialist Linda Young, in her review of *Handbook of Material Culture* provides comment on a section of the book that deals with the body, materiality and the senses. Young says: "...[the somatic] confronts textuality and visuality as our culture's dominant modes of understanding material culture, and suggests that the embodied subject and its multiple, concomitant ways of sensing, feeling, knowing, performing and experiencing, offer dynamic routes to different perceptions of the human relation to the material... Corporeality and sensuality open up to the concept of sense-scapes—an enticing notion."[21]

[20] Ihde, D. 'Material hermeneutics', in *Symmetrical archaeology*, Stanford: Theoretical Archaeology Group (2005), available online <http://humanitieslab.stanford.edu/Symmetry/746>, last accessed 3 August 2007.

[21] Young, L. review of *Handbook of material culture* (2007) Tilley, C., Keane, W., Kuechler, S., Rowlands, M. & Spyer, P. eds. (2006), in *reCollections: Journal of the National Museum of Australia*, vol. 2, no. 2, available online <http://recollections.nma.gov.au/issues/vol_2_no2/book_reviews/handbook_of_material_culture/>, last accessed 30 June 2009.

2 PLACE-HAMPI

Figure 1. PLACE-Hampi installation at Lille Opera House, 2006.

PLACE-Hampi[22] is a vibrant theatre for embodied participation in the drama of Hindu mythology focused at the most significant archaeological, historical and sacred locations of the World Heritage site Vijayanagara (Hampi), South India.[23] The work's aesthetic and representational features constitute a new approach to the rendering of cultural experience and give viewers a dramatic appreciation of the many-layered significations of this site (fig. 1).

PLACE-Hampi was developed in the context of an Australian Research Council Linkage Grant 'Reformulating narrative in virtual heritage, using a co-evolutionary model of immersive interactivity'.[24] The work has been described in a number of scientific papers[25] however its

[22] PLACE-Hampi <http://place-hampi.museum> last accessed 30 June 2009. This work was commissioned for Lille3000 in 2006. It has been exhibited at Lille3000, Opera House, France Oct 2006-Jan 2007; Berlin Festival, Von Funken Zum Pixel, Kunst + Neue Medien, Martin Gropius Bau, Oct 2007-Jan 2008; Panorama Festival, KZM, Karlsruhe, Jan 2008-Mar 2008; eArts/eLandscapes, Museum Science & Technology, Shanghai Oct 2008; Ancient Hampi: The Hindu Kingdom Brought to Life, Museum Victoria, Nov 2008-Jan 2010.

[23] In recognition of the significance of Vijayanagara, the 'Hampi Group of Monuments' was inscribed on UNESCO's World Heritage List in 1987. For a comprehensive discussion of the archaeology and history, see Vijayanagara Research Project <http://www.vijayanagara.org/>, last accessed 30 June 2009.

[24] Australian Research Council Linkage Grant 2006-2009: 'Reformulating narrative in virtual heritage using a co-evolutionary model of immersive interactivity' (LP0669163).

[25] Kenderdine, S., Shaw, J., Del Favero, D. & Brown, E., 'Place-Hampi: Co-evolutionary narrative & augmented stereographic panoramas, Vijayanagara, India', in Kalay, Y., Kvan, T. & Affleck, J. (eds), *New heritage: New media and cultural heritage* (Abingdon: Routledge/Taylor & Francis Books 2007), 336–52.
Kenderdine, S. 'Somatic solidarity, magical realism and animating popular gods: Place-Hampi "where intensities are felt"', *Proceedings of the 11th European Information Visualisation Conference*, IV07, Zürich, July 2007, IEEE Comp Society (2007a): 402–8.
Kenderdine, S. 'The irreducible ensemble: Place-Hampi', *Proceedings of Virtual Systems and Multimedia 13th International Conference*, VSMM 2007, Brisbane, 23–26 September 2007, Revised Selected Papers Series: Lecture Notes in Computer Science, Subseries: Information Systems and Applications, incl. Internet/Web, and HCI, vol. 4820, Berlin: Springer, (2007b): 58–67.
Kenderdine, S., Shaw, J. & Kocsis, A. 'Dramaturgies of PLACE: Evaluation, embodiment and performance in PLACE-Hampi', DIMEA/ACE Conference (5th Advances in Computer Entertainment Technology Conference & 3rd Digital Interactive Media Entertainment and Arts Conference), Athens,

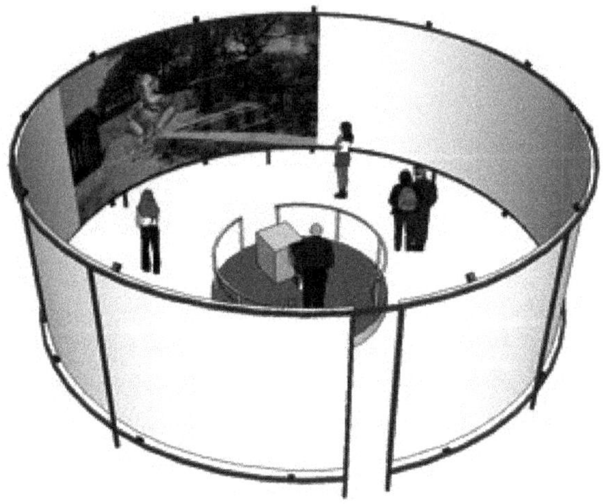

Figure 2. Graphic representation of PLACE-Hampi.

Figure 3. PLACE-Hampi, LCD navigation interface and panoramic cylinders.

basic attributes will be included here for the clarity of the reader.

PLACE-Hampi uses an innovative visualization environment based on the artistic paradigm invented by Jeffrey Shaw in 1995 for PLACE-A User's Manual[26] (redesigned of PLACE-Hampi by Kenderdine and Shaw in 2006). Its interaction scenario is based on is a panoramic 360-degree screen in the centre of which is a large motorized platform that allows the viewer to rotate the projected image around its cylindrical screen (fig. 2). Standing on the platform the viewer can control the rotation of the platform to the left and to the right, which in turn rotates the projection system that is also mounted on this platform. Beside interface's LCD screen are buttons that allow the viewers to also control their forwards and backwards movements within the 3D world on the screen (fig. 3).

For the PLACE-Hampi implementation, stereoscopic 3D projection was used for the first time, and a key feature of this work is its innovative use of stereoscopic 3D panoramic photography (fig. 4). A custom built pair of Seitz Roundshot 220 format film cameras were used to shoot two panoramas simultaneously, one for the left eye and the other for the right as it rotates 360 degrees while recording.

Over one hundred panoramic pairs were shot in Hampi, and eighteen were selected for the final work showing many of its most significant sites. These locations include important contemporary pilgrimage sites and temple precincts associated with the epic poem Ramayana, many of which encompass spectacular views over the Hampi landscape.[27]

Figure 4. Seitz Stereo VR Roundshot camera.

The virtual setting of PLACE-Hampi is a representative boulder strewn landscape dominating the surrounding horizon. This landscape encloses the constellation of eighteen cylinders. The ground surface of the landscape is inscribed with an iconic drawing of the simian god Hanuman, a symbolic reference to Kishkindha, which is the mythological monkey kingdom depicted in the Ramayana and considered by many Hindus to be located at Hampi. The LCD monitor on the interaction platform shows

November 2009 (2009) ACM forthcoming.

26 See PLACE-A Users Manual, available online <http://www.jeffrey-shaw.net/html_main/frameset-works.php3>, last accessed 30 June 2009.

27 These locations are described in the interactive map on the PLACE-Hampi website. Available online <http://place-hampi.museum/contents/map.html> last accessed 30 June 2009.

Figure 5. Motion capture of an Indian classical dancer, performance of Shiva as Nataraja.

Figure 6. Corresponding animation of Shiva as Nataraja.

an aerial view of this allegorical map and the eighteen panoramic cylinders are strategically positioned on this ground plan of Hanuman's body. The map also locates the viewer's real time position and movements within this imaginary/documentary terrain, and he/she can virtually travel in any direction, encountering and entering the cylindrical panoramas that endlessly replicate themselves in every direction to an infinite horizon.

Embedded within the rich scenery of four of these panoramas, and composited into their three dimensional landscapes, are lively narrative events enacted by computer graphic characters based on Hindu mythologies, including Ganesha, Lord Shiva and Garuda (figs 5 & 6). The aesthetic is based on India Magical Realism and popular contemporary Hindu representation of the Gods appropriate to the place.[28] A further augmentation is the stylized three-dimensional Sanskrit texts that can also be released into the virtual Hampi landscape via a button on the user interface. These computer-generated texts are citations from chapter 13 and 37 of the Ramayana, and describe the gathering of the monkeys at Kishkindha. Letters, words and sentences follow the trajectory of the viewer's movements through the landscape and create a visual tracery of narrative paths that evidence the active presence of its visitors.

The visual landscape is conjoined by a spatial aural field made from ambisonic 360-degree recordings that were captured in situ at Hampi, at the exact location and time of each panoramic photographs (fig. 7). These provide a temporal component to the photographic stills. The selection of specific panoramas that are augmented with animations of Hindu gods are additionally enlivened by classical Carnatic musical compositions selected and composed by the eminent Indian Carnatic violinist Dr L Subramanian. As the visitor navigates PLACE-Hampi, the conjunction of these singular audio, visual and interactive strategies of representation articulate a high level of viewer co-presence in the narrative exploration of a virtual cultural landscape.

Figure 7. Ambisonic field recordings at Hampi.

[28] The animations were developed by artists and technicians at the Indian animation house Papriakaas. For extensive discussion of the basis for animation in relation to darshan, and Christopher Pinney's notion of corpothetics and magical realism, see Kenderdine 2007a, op cit.

While PLACE-Hampi embodies a single user interaction paradigm, the emergent narrative relations that enliven its panoramic scenes with mythological significance constitute a performance space that kinesthetically involves the entire audience. Visitors standing within the installation's panoramic enclosure find themselves walking around to follow the circular path of the projected image as it is rotated around the screen by the operator. Their physical action walking about in the real space has proprioceptive concurrency with their movements within the virtual space. This corporeal conjunction of real and virtual modalities of passage, reinforced by the real and virtual/stereoscopic modalities of 3D representation of place, constitutes for almost all viewer's an intense experience of immersion, presence and engagement in the work's aesthetic and conceptual constructs. This was made evident in extensive user evaluations carried out at its exhibition in Berlin (Martin Gropius Bau) and Melbourne (Immigration Museum), which focused the phenomenological and experiential analysis of user experience, performance issues of engaging with the work and on cross-cultural understandings.

2.1 Evaluating PLACE-Hampi

The analysis of the PLACE-Hampi installation has provided rich observational and quantitative data on the power of stereoscopic, panoramic interactive display systems for presence in the exploration of heritage landscapes. The research using the results of a four-page form based questionnaire[29] through a phenomenological [32] framework aimed to draw specific insights and enrich the existing discussion of audience experience inside the immersive interactive works. The results of the analysis are highly significant for designers of situated multimodal immersive entertainment in museums and galleries and have been discussed in a prior research paper to which the reader is referred.[30] For the purposes of this paper, it is worthwhile to reinforce the needs of extensive and in-depth evaluation for practitioners in the field.

The evaluation of PLACE-Hampi fills a useful gap in knowledge for those working in interpretive heritage, museums and other cultural agencies for three primary reasons. Firstly, there are few evaluations done on cultural heritage based virtual reality works in the public domain (notable exceptions are the research of heritage and museum professionals working with technologies such as Dallas[31] Forte *et al*[32] Economou & Pujol[33] Bonini,[34] Roussou[35]). Prior research in the field of virtual heritage applications has pointed out "... a wide percentage of projects and applications of virtual heritage are never experimented and monitored with people, but they born and die in digital labs".[36] In addition, evaluation of new media and virtual heritage applications has been problematic, without any well-defined guidelines for conducting such assessments. In many cases, the evaluation is related to the content itself ("was it an accurate 3D model or not?") and not the analysis of users' interactions and behaviours.[37]

Secondly, "interaction and feedback determine the virtual embodiment"[38] a vital component to "the empathy factor really crucial for learning and communication."[39] The PLACE-Hampi survey provides many examples that explicitly draw out this interaction paradigm. The qualitative findings support the strategy that multimodal interactive worlds that focus on kinaesthetic and multisensory amplification can play a significant role in the interpretation heritage landscapes. As post-processual archaeology focuses on media, the importance of embodiment and 'presence' and, an artistically informed inquiry into interpretation comes strikingly to the fore.

For example, see post-processual theorists Webmoor,[40]

---

[29] Aggregated data for 284 respondents for the evaluation survey undertaken at Martin Gropius Bau, Berlin 2007 <http://place-hampi.museum/pdf/Avie_Berlin_Questionnaire_08_data_screen.pdf>,last accessed 30 June 2009.

[30] For extensive discussion, see Kenderdine, et al 2009 op cit.

[31] Dallas, C., 'The presence of visitors in virtual museum exhibitions'. Numérisation, lien social, lectures colloquium, University of Crete, Rethymnon, 3–4 June. Available online at Museology, Issue 4: Performitivity, Interactivity, Virtuality, and Museums, (2004), available online <http://museology.ct.aegean.gr/indexEE.php>, last accessed Friday, April 10, 2009.

[32] Forte, M., Pescarin, S. & Pujol, L. 'VR applications, new devices and museums: visitors' feedback and learning. A preliminary report'. The e-volution of Information Technology in Cultural Heritage. Where Hi-Tech Touches the Past: Risks and Challenges for the 21st century, *Proceedings of the 7th International Symposium on Virtual Reality, Archaeology and Cultural Heritage*, VAST 2006, short papers volume, Eurographics, (2006): 64-69.

[33] Economou, M. & Pujol, L. 'Educational tool or expensive toy? Evaluating VR evaluation and its relevance for virtual heritage', in Kalay, Y., Kvan, T. & Affleck, J. (eds), *New heritage: New media and cultural heritage*, (Abingdon: Routledge/Taylor & Francis Books 2007), 284–302. Economou, M. & Pujol, L. 'Exploring the suitability of virtual reality interactivity for exhibitions through an integrated evaluation: the case of the Ename Museum', *International Museology Journal*, Department of Cultural Technology and Communication, University of the Aegean, vol. 4, (2007) available online <http://museology.ct.aegean.gr/articles/2007125113247.pdf>, last accessed 30 June 2009.

[34] Bonini, E. 2008, 'Building virtual cultural heritage environments: the embodied mind at the core of the learning processes', *International Journal of Digital Culture and Electronic Tourism* 2: 2/2, (2008): 113-125.

[35] Roussou, M. 'Virtual heritage: from the research lab to the broad public, Virtual Archaeology, *Proceedings of the VAST Euroconference*, Arezzo 24-25 November, (2000) 93-100.

[36] Forte, M., Pescarin, S. & Pujol, L. (2006) op cit: 68.

[37] Ibid. 64-69.

[38] Ibid. 68

[39] Forte, M., Pescarin, S., Pietroni, E. & Rufa, C., 'Multiuser interaction in an archaeological landscape: The Flaminia project', in Forte, M. & Campana, S. (eds.), *From space to place, Proceedings of the 2nd International Conference on Remote Sensing in Archaeology*, Rome, 4–7 December 2006, British Archaeological Reports International Series 1568, Oxford: BAR Publishing, (2006): 189–96.

[40] Webmoor, T. 2007, 'Lessons From the Real: Mediating People-things in a Symmetrical Archaeology' (2007), available online <http://traumwerk.stanford.edu:3455/Symmetry/743> last accessed 30 June 2009. Webmoor, T., 'Mediational techniques and conceptual frameworks in archaeology', *Journal of Social Archaeology*, Vol. 5, No. 1, (2005): 54–86.

Witmore[41] and Shanks.[42] The Presence Project (a conjunction of scientists, archaeologists and artists) recognizes the importance of immersion to this cyber theatre of archaeology.[43]

Finally, cyber archaeology researchers Forte and Bonini reinforce the necessity for considering the cognitive learning models through *enaction* and embodiment emphasizing that "knowledge is enfolded in movement".[44] This is the premise for body-anchored and experience based learning. Similar arguments for a primary acknowledgement of the body (and cognition) in the creation of meaning have been occurring across material studies, new media and cinema studies for some time.[45] Place-Hampi, supported by its analysis, extends this to the domains of interpretive cultural heritage and asserts a primacy for enaction through its strategies of immersive/ interactive architecture and content development.

PLACE-Hampi's conjunction of stereo-vision (via its stereoscopic renderings of the Hampi landscape) and optic flow (via the mobility of the viewer's bodies within the projection arena) allows it to constitute a level of perceptual and kinaesthetic realism that explains the strong engagement of its virtual world that the majority of viewer's expressed in the evaluation reports. Users of the PLACE-Hampi system can be summarized by the following comment: "I myself am part of *PLACE-Hampi* and I determine in which part of the artwork I stay".[46]

Respondents to the Martin Gropius Bau survey registered the impacts on their bodies by indicating upon a blank diagram of the body. Perhaps predictably due to the stereoscopic and ambisonic spatialized audio strategies in the application, the emphasis for engagement with the body focused prominently on the eyes, ears and hands (and is related to driving the rotating platform). For the aggregated data from all 284 respondents in a single diagram see Figure 8 and for examples of individual respondents see Figure 9.

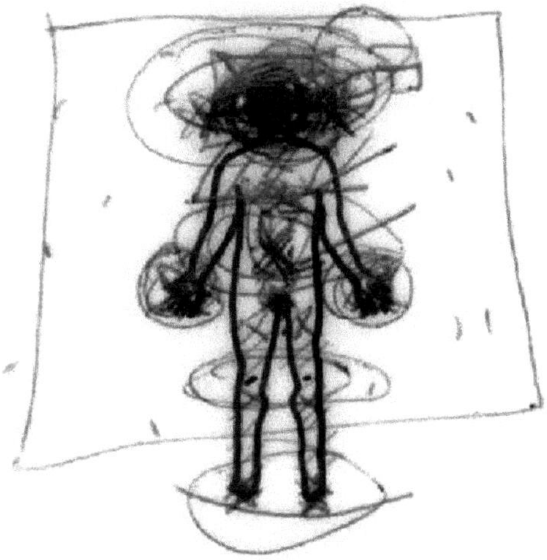

Figure 8. Aggregated data from the evaluation survey, Martin Gropius Bau, Berlin, 2007.

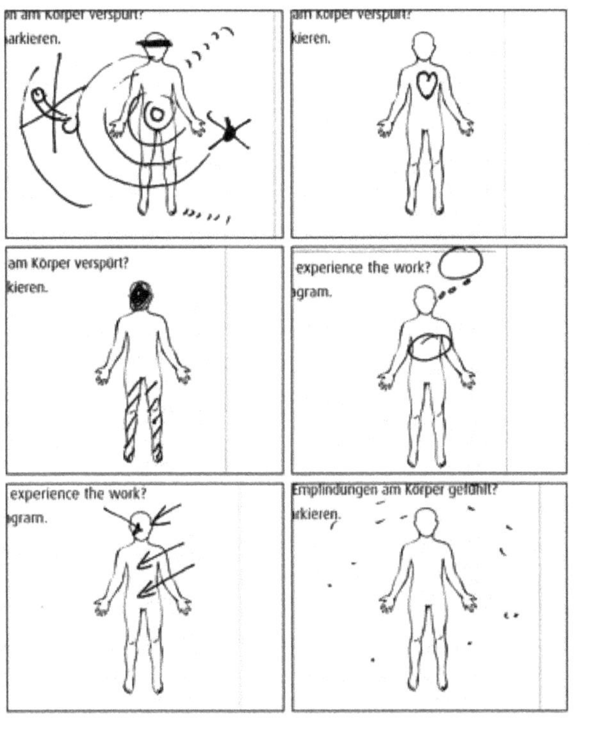

Figure 9. Six responses indicating experiential phenomena, from the evaluation survey, Martin Gropius Bau, Berlin, 2007.

It is worthwhile to note here that several responses to this question also indicated that physical responses were established beyond the 'mere' cognitive functions of viewing and steering the platform, and point to more 'esoteric' physical and indeed properly 'embodied' connections with the work (fig 9).

The reader is referred to Kenderdine, Shaw & Kocsis (2009) for a greater analysis. The research validates arguments made earlier in this paperfor the privileging of the embodied world of interpretation.

---

[41] Witmore, C.L., 'Vision, media, noise and the percolation of time: symmetrical approaches to the mediation of the material world', *Journal of Material Culture*, Vol. 11, No. 3, (2006): 267–292.
Witmore, C.L., 'Four archaeological engagements with place: mediating bodily experience through peripatetic video, The Visual Anthropology Review, Vol. 20, No. 2, (2004): 57–72.

[42] Shanks, M., 2006. Media and archaeological futures DOI= http://documents.stanford.edu/michaelshanks/113 and Pearson, M. & Shanks, M. 2001*, Theatre/Archaeology*, London: Routledge.

[43] The Presence Project, available online (2008) <http://presence.stanford.edu:3455/Collaboratory/19>, last accessed 30 June 2009.

[44] Bonini, E. 2008, op cit. 123.

[45] Pinney, C. 'Four types of visual culture', in Tilley, C., Keane, W., Kuechler, S., Rowlands, M. & Spyer, P. (eds), *Handbook of material culture,* London: Sage (2006): 131–44.
Shoback, V. *Carnal Thoughts: Embodiment and Moving Image Culture* (University of California Press 2004).
Stafford, B.M. The Remaining 10%: The Role of Sensory Knowledge in the Age of the Self-Organizing Brain. Visual Literacy, (ed.) J. Elkins (New York, Routledge 2005).

[46] Kenderdine, S. & Schettino, P. 'PLACE-Hampi: interactive cinema & new narratives of embodied cultural experience', *Inclusive Museum conference*, Istanbul June 2010. (2010) Abstract submitted.

2.2 Cyber/Theatre/ Archaeology

Digital technologies can be contextualized within the historical frameworks of human experience and immersion in all types of media. Interactive and immersive cinema has clear links to performance, ritual, theatre, and the circus. The inter-play between the immersion and the nature of the interactive cinematic of PLACE-Hampi is highlighted through such comments as: "[...] one gets the feeling that one is 'inside' the film and can direct the film"[47] and, "It has something of a stroll through a virtual world and I am my own cameraman".[48]

Recognizing such performative qualities of the human-computer interface,[49] McKenzie suggested 'one might invent the computer as performance'.[50] Media theorist Gabriella Gianacchi, in her analysis of the virtual theatre, describes it as: "...one which through its virtuality is able not only to include the viewers within the art but also to distribute their presence globally in both the real and simulated virtual world."[51]

The relationships between theatre, performance and archaeology also provide a framework for the (re) articulation of fragments of the past as real-time events. In the emergent, narrative encounters of the installations described in this paper a new staging takes place in these immersive settings. This staging is the encounter between the user of the system, the materiality of landscape and archaeological, historical and intangible knowledge. This encounter is performed in the present between the virtual and the real, between human and machine agent. PLACE-Hampi is analyzed here as a case study looking at attributes of 'cyber theatre' and as an example of the resocialization of public space[52] in combination with the mediation of an archaeological landscape.

In 2001, archaeologist Michael Shanks and performance studies theorist Mike Pearson collaborated to write *Theatre/Archaeology*[53], an examination of the points of convergence between contemporary performance theory and practice and interpretive approaches in archaeology. Invoking the notions of performance in archaeology helped to make sense of the matrix of associations between archaeological materials—empirical, spatial, conceptual and metaphorical. For Shanks, archaeological interpretation should also emphasize *manifestation* of archaeological materials, which is "letting the material display itself."[54] He also notes the emphasis of performance, ceremony and ritual in relation to post-processual archaeology. This ties together "issues of signification, of the embodiment and corporeality of social actors, agency and the constitution of social structure and social norms."[55] Shanks concludes, "performance is the root metaphor for social and cultural processes"[56] and further, archaeology itself can be conceived of as a performance "where the remains of the past are mobilized in practice, often conceived as mimetic, of representing or restoring behaviour."[57]

These ideas can be extrapolated directly to new media theories of performance and spectatorship, drawing upon the studies of PLACE-Hampi as an embodied theatre of participation. From the perspective of the social interaction and individual/group interaction within PLACE-Hampi, it is worthwhile to explore the dynamic series of relationships as *performance* in this cybernetic theatre. As art historian and media theorist Ron Burnett points out "Digital technologies need to be contextualized within the historical frameworks of human experience and immersion in all types of media and interactive and immersive cinema has clear links to performance, ritual, theatre, painting, the circus and painting".[58]

The theoretical discussion of performative qualities of the virtual theatre often neglects the primary communication that occurs between people in the real-world space as they perform the act of spectatorship or user participation. The aesthetics of interaction is "rooted in the user's experience of herself performing her perception."[59] Both performance theory and sociology, when considering how a Human Computer Interface (HCI) works, suggest that the user is simultaneously the *operator* of the system, the *performer* of the system and the *spectator*. In multi-participatory works, which embody a single-operator/ user and multiple spectators, numerous bonds exist between the user and the spectators, and the user and the system.

Between the user and the system, the concept of embodiment is of primary concern. Embodiment is a 'participatory' status and a foundation for exploring interaction in context.[60] In terms of the *trichotomy* of the system-user-spectators, embodiment implies a reciprocal relationship with the context, encompassing users, interactive systems, spectators, co-users, physical surroundings, and the meanings ascribed to these

[47] Ibid.
[48] Ibid.
[49] Brenda Laurel (1993) wrote a seminal work on computers as theatres that set the stage for the discussions that followed.
[50] McKenzie J.'Virtual Reality: Performance, Immersion, and the Thaw', TDR: The Drama Review 38.4 (1994): 83-106. Reprinted in Performance: Critical Concepts in Literary and Cultural Studies, Volume 1. Philip Auslander, ed. (London: Routledge 2004).
[51] Giannachi G. *Virtual theatres: An introduction*, (London: Routledge 2004).
[52] A primary motivation has been to animate a resocialization of public spaces, reinforced by the understanding of museums as places of 'civic seeing' and 'zones of contact' for people (Bennett 2006: 263–281). Situated, immersive experiences bring people together in real spaces to interact not only with the virtual 'other' and the virtual objects of encounter but also amongst themselves.
[53] Pearson & Shanks, 2001 op cit.

[54] Shanks, M. 'Three rooms: Archaeology and performance', in *Journal of Social Archaeology*, vol. 4, no. 2, (2004): 147–80, 148.
[55] Ibid.149.
[56] Ibid.
[57] Ibid.
[58] Burnett R. *How images think*, (Cambridge, MA, London: MIT Press 2005), 129
[59] Dalsgaard, P. & Koefoed-Hansen, L. 'Performing perception—staging aesthetics of interaction', ACM *Transactions on Computer-Human Interaction* (TOCHI), Vol. 15, issue 3, Article No. 13: (2008): 1.
[60] Dourish, P. 'Seeking a foundation for context-aware computing', *Human-Computer Interaction*, vol. 16, no. 2, (2001): 229–41.

entities.[61] Researchers of computer-human interaction Reeves, Benford, O'Malley and Fraser[62] have addressed the issue of how a spectator should experience a user's interaction with the computer.[63] Borrowing from performance theory, the user is the *inter-actor* with the system and the interaction between the user and the system is the *performance*. While this relationship is what is mostly described in media art and HCI, it is the *spectators'* relation to and experience of the performance that is also of interest here. This aspect is particularly relevant if we consider the resocialization of public spaces using digital technology and through immersive installations situated in museums.[64] In addition, as Dalsgaard and Koefoed-Hansen describe: "It is the ways in which the user perceives and experiences the act of interacting with the system under the potential scrutiny of spectators that greatly influences the interaction as a whole... it is precisely this awareness of the (potentiality of a) spectator that transforms the user into a performer".[65]

The key to this relationship is the 'awareness' of others, which provides the context for individual activity. The tension that occurs is between the spectators watching the user and the user's awareness of being the centre of the spectators' gaze. The user not only acts in relation to the system but also is propelled by the knowledge that her perception of the system is a performance for others. Dalsgaard and Koefoed-Hansen call this *performing perception*.[66] The user simultaneously engages in three actions: the act of interacting with the system; the act of perceiving herself in relation to the system and her surroundings; and the act of performing.[67] The spectators invest in the user as surrogate selves, demanding a 'correct' performance of the system. This interchange between all participants becomes a defining attribute, giving rise to meanings of the social for interaction design. A portion of the PLACE-Hampi survey in Berlin focused on the performative aspects of the trichotomic relationships (user-system-spectator), with visitors being asked, 'was it a social experience that you shared with other people?'; 137 of the responses confirmed this. Indeed, it is frequently observed that spectators will 'direct' the user of the system, indicating their preference of where to go in the virtual world. In addition, spectators often encourage the user to relinquish control if they themselves feel eager to direct the journey.

Single-user interfaces such as PLACE-Hampi will always engender this inherent tension of negotiated use; however, the social dynamics it sets up are one of its desirable features. Co-experience is understood as the experience that users themselves create in social interaction and which is shared with others. Co-presence, the tacit awareness of others' presence in the space, is evident in the PLACE-Hampi data. The proximity of other audience members and the continual, rotational movement of the platform predicate an awareness of dwelling together, accommodating, relocating and finding space. This collaborative journeying in PLACE-Hampi bears the cognitive challenges of inhabiting the virtual and real worlds simultaneously.

In the analysis of PLACE-Hampi, it is clear that visitors felt virtually no trepidation or hesitation in using the interface. Only seven of the 284 respondents noted that they did not 'feel confident' while 117 respondents used the interface platform without hesitation.[68] The vast majority of interface users considered 'it easy to use' (119 'yes'; 7 'no'). For a museum environment, this level of visitor interaction presents an impressive outcome. It is worth pointing out too that the collective approach to understanding an interface (one with no instructions) creates a social dynamic of exploration and self-discovery/learning that is compromised if delivered by a customer service manager or dedicated guide. The design philosophy of PLACE-Hampi is that the system should act independently of expert interpretation and be available for non-specialist discovery.

While a majority of respondents indicated that it was more enjoyable to 'drive' than to 'watch' (84) a significant number preferred watching (38). Driving and watching, it seems, are both pleasurable. Participants then reflected on the question, 'were you self-conscious that other people were watching you [while operating the interface]?' Ninety respondents said 'yes', 34 'no'. Perhaps for this reason, simply watching is a pleasure in the performance of journeying through and around PLACE-Hampi. Furthermore, the 38 responses in favour of watching are open to psychological interpretations. Firstly, the preference could be related to subjective comfort as it presents a safe mode of being inside PLACE-Hampi, a kind of withdrawal from the more active role of 'driving'. Secondly, the preference may be related to enjoying the visual display and may in fact constitute a positive *scopophilic* dimension, 'the love of watching'. Most people have a history of watching that comes from the cinema or TV, rather than one of interaction and performance in immersive installations. These are some of the dramaturgies of PLACE-Hampi.

2.3 YER-Türkey

PLACE-Hampi and its evaluation forms the basis for YER-Türkey, comprised of thirty five historic and contemporary sites throughout Turkey documented through stereographic panoramas and, four monoscopic video based panoramic movies (captured using the

[61] Dalsgaard & Koefoed-Hansen 2008, op cit. cf Dourish (2001), 5.

[62] Reeves, S. Benford, S, O'Malley, C. and Fraser, M. Designing the spectator experience, *Proceedings of the Conference of Human factors in Computer Systems* (CHI05), ACM, (2005): 741-750, 748

[63] Ibid.

[64] Steve Benford and his associates at Collaborative Computing in the Mixed Reality Laboratory at the University of Nottingham extend the user-spectator relation through a series of locative media interactive game/ performances (Benford *et al.* 2006). *Uncle Roy All Around You*, available online <http://www.mrl.nott.ac.uk/~sdb/research/downloadable%20 papers/frame%20of%20the%20game.pdf>, last accessed 25 May 2009.

[65] Dalsgaard & Koefoed-Hansen 2008, op cit. 6.

[66] Ibid. 31.

[67] Ibid.

[68] See Kenderdine, et al 2009, op cit.

LadyBug®3) capturing intangible heritage and psycho geographic interpretations of place. This work is currently in production for the Borusan Music House[69], Istanbul (premier June 2010).

3. Ancient Hampi

Ancient Hampi: the Hindi Kingdom Brought to Life is an exhibition without objects that seeks to engage a multi-cultrual public in phenomenological, sensorial and experiential encounters with an archaeological precinct, historic place and a living cultural landscape. This exhibition draws upon the products of the archaeological imagination and the visual languages of archaeology, photography and new media converge to reveal both the ancient Hindu kingdom of Vijayanagara and Hampi, as a living community.

The exhibition was designed in three parts to contextualize the interactive and immersive installation PLACE-Hampi at the Immigration Museum, Melbourne, Australia (2008-2010). These parts are: Archaeology, Photography and the aforementioned PLACE-Hampi.

3.1 Archaeology Room

Archaeology is "a discipline of things"[70] and the challenge then, in the absence of objects, is how to acknowledge the richness of this ancient place, far from the tangible site itself. Archaeology may be about describing relationships from the past to the present. The discipline is defined by not only an efflorescence of words but also the techniques of visualization and mapping, and its tools include everything from measuring tapes to Global Information Systems.

Drawing from the extensive archive of archaeologist Dr John Fritz and art historian Dr George Michell, Ancient Hampi provides the opportunity to discover the fascinating contours of their archaeological practice. Through an interactive light table and interviews, this exhibition offers a display of archaeological processes often unseen in the final rendering of scholarly and popular publications.

The Archaeology Room features three components, including an interactive light table[71] that invites visitors to pick any of the 120 images (printed to A5 transparencies) and scan the attached barcode, to get a high resolution image projected on the wall and an associated scholarly label of what is depicted in the image. The images comprise of photographs from the 30 years of work at the site by the archaeologists and their colleagues including; site plans heavily annotated with corrections, drawings and renderings of features, and fieldwork notebooks. This

Figures 10 & 11. Interactive archaeological light table, Ancient Hampi exhibition.

interactive feature is a way to 'rummage' in the archive of materials that do not usually find their way into the pristine pages of publications (figs 10 & 11).

In the Archaeology Room, there is also a video conversation with the two archaeologists (Fritz and Michell, Kenderdine,) taken in the Photography Room of the exhibition, just before opening with the complete installation in situ. This interview was taken using a panoramic video camera (LadyBug®2) and the resultant panoramic movie is rendered as a cylinder that slowly rotates, so that the interviewer and interviewees are conjoined in conversation (fig 12). This unexpected viewing conjuncture provides a fascinating alternative to traditional interview techniques. The conversations that take place also resonate with the user of interactive light table providing an obviously contemporary commentary for the archival visual materials of the transparencies.

An additional element in this room is a kiosk with the PLACE-Hampi website[72] which contains the didactic information to supplement an otherwise experiential exhibition. It also contains an interactive map[73] of the

[69] Borusan Music House, available online < http://www.borusansanat.com>

[70] Olsen 2003, cited in Webmoor, T. & Witmore, C. *Symmetrical archaeology*, (2005) available online <http://traumwerk.stanford.edu:3455/Symmetry/home>, last accessed 30 June 2009.

[71] This light table draws inspiration from artist Tjebbe van Tijen's *SITUATIONISTEN OPDRIFT 2D*, (2006). Tjebbe acted as a consultant to Ancient Hampi. See Imaginary Museum Projects, available online <http://www.imaginarymuseum.org/ >, last accessed 30th June 2009.

[72] PLACE-Hampi website <http://place-hampi.museum> and particularly Ancient Hampi exhibition <http://www.place-hampi.museum/contents/exhibition.html>, last accessed 30 June 2009.

[73] PLACE-Hampi interactive map <http://www.place-hampi.museum/

Figure 12. Panoramic interview, Ancient Hampi exhibition.

Figure 13. Photography Room, Ancient Hampi exhibition. Photo: Ben Healley

panoramic locations in PLACE-Hampi, with location and historical information on each scene.

3.2 Photography Room

Photography provides archaeology with a potent method of translation and representation of the visible remains of the past. Photography also gives us new ways of looking at cultural landscapes and establishes a sense of place.

Drawing in part from the vast collection of renowned architectural photographer John Gollings, who has documented the site over the last 25 years, Ancient Hampi presents, for the first time, a myriad of impressions from its extraordinary geography and architecture to the vibrancy of its social life. Along the dark corridor entrance to the Photography Room, projected at a scale of 2.5 metres across are a series of black and white night shots taken at Hampi of the ruins.

Figure 14. Magnifying Hampi, school activities, wAncient Hampi exhibition. Photo: Ben Healley.

Inside the Photography Room fifty photographic images by Gollings of architectural features at Hampi, each one-metre wide are double-hung around this room provide a walk-in image landscape (fig 13). In the middle of the room, there is a 9mx600mm light table showing all the left-eye of the stereo panoramic images from the PLACE-Hampi shoot (many more than appear in PLACE-Hampi itself), together with fieldwork shots taken during the creation of PLACE-Hampi. Five large magnifying glasses allow visitors to scrutinize the details on these high-resolution images (fig. 14).

The Exhibition has become a major focus for the Hindu community in Melbourne and extensive qualitative evaluation of the cross-cultural impacts of the PLACE-Hampi system within the greater context of the Exhibition is currently underway.[74]

4. Hampi-LIVE

Hampi-LIVE is the second part of the ARC Linkage Grant 'Reformulating narrative in virtual heritage, using a co-evolutionary model of immersive interactivity'.[75] The work exploits the technological and aesthetic features of UNSW iCinema Centre's Advanced Visualization and Interaction Environment (AVIE, fig. 15) including camera tracking, omni-stereoscopic projection and multi-channel spatialized sound. The cylindrical AVIE screen is 10 metres in diameter and 4 metres high.[76] 12 SXGA+ DLP projectors are mounted in polarized stereo pairs to illuminate the entire cylinder, resulting in a total circumferential resolution of ~ 8000 pixels and vertical

contents/map.html>, last accessed 30 June 2009.

[74] See Kenderdine & Schettino 2010 op cit.

[75] Australian Research Council Linkage Grant 2006-2009: 'Reformulating narrative in virtual heritage using a co-evolutionary model of immersive interactivity' (LP0669163).

[76] Advanced Visualisation and Interaction Environment <http://icinema.unsw.edu.au/projects/infra_avie.html>, last accessed 30 June 2009.

resolution of ~ 1000 pixels. A cluster of 6 dual Xeon Windows PC's is used to drive the projectors, and the system has 18 speakers distributed around and outside the acoustically perforated screen. Five overhead infrared cameras provide coverage of the entire AVIE arena, and infrared floodlights provide the illumination. From this camera data the AVIE system can dynamically track the viewers' positions and movements within its projection space. As stereoscopy is a powerful means of enhancing immersion, AVIE's omnistereo rendering algorithm assumes a viewpoint at the centre of the cylinder and a view direction perpendicular to the screen surface. This method produces perceptually correct stereoscopic depth over the full 360º viewing circle and provides all viewers a valid stereo image.[77]

With these facilities, Hampi-LIVE augments the original stereoscopic panoramic content of PLACE-Hampi that is navigable in AVIE using a custom joystick positioned in the centre of the projection cylinder. Now it becomes a 360-degree context to explore the potentialities of co-evolutionary narrative whereby audience behaviour can influence and be influenced by the real-time behaviour of virtual characters (machine agents) in a mixed reality setting. This experimental research is embodied in two scenarios that exploits AVIE's infrared camera tracking to enable an operational and proprioceptive interface between the real and virtual worlds.

Presence flourishes within immersive environments in which the behaviour of virtual characters can co-evolve interactively, by making intelligent references to the actions of viewing participants in real-time.[78] When these co-evolving systems of interaction are applied, unique relationships are formed between viewers and computer-generated characters - often in dramatic, culturally distinctive ways. This facilitates dynamic inter-actor participation, cultural learning, and the creation of presence in virtual heritage.

In this scenario, real time animated monkeys populate one of the 3D photographic panoramas shot at Hampi - specifically, a site located on Hemakuta Hill behind the main active temple at Hampi. Referencing a 3D database that describes the physical layout of this panorama, these monkey agents coherently and seemingly intelligently move about it's architectural and geographical features. Taking information from the camera tracking system they can also seemingly sense where the visitors are in the AVIE as well as follow their movements. In conjunction with the camera tracking system, their behavioural scripts embody a co-evolutionary relationship to the behaviours of the real visitors within AVIE.

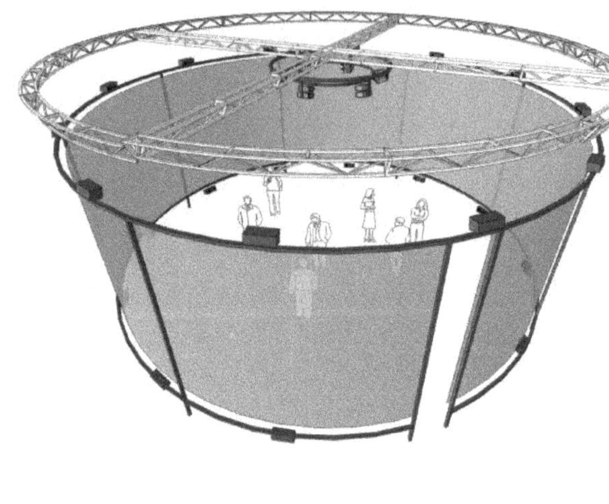

Figure 15. AVIE © iCinema Centre, UNSW.

Monkeys are an ubiquitous presence at Hampi. In this work, their identity and aesthetic articulation is closely connected to Hampi's identification as Kishkindha, the kingdom of the monkeys (figs 16-17). The mythological inhabitants have counterparts in the real world, and at Hampi today, the prevalent monkeys are revered by the faithful yet are often delinquent in their behaviour towards permanent inhabitants and tourists alike (fig. 18).

The evolving time-based development of behavioural interaction between real people and virtual agents is achieved by script-based actions of computer graphic animations constituted from motion-captured sequences and linked to interpretive software techniques and high-level computer programming (figs 19 & 20).

The software allows the machine agents to act, observe the consequences of their actions in the real world, and then formulate new actions according to a library of scripts and improvisational capabilities that inform their individual identities. For example, a mother monkey may prioritize the protection of her young and will take appropriate action to protect her territory from the proximity of humans. Other monkeys are given various drives towards socialization with the human visitors (e.g. hunger for food, interest in bodily antics, or mere curiosity). Different temperaments can be defined, such as fearless, jittery, protective, paranoid, etc. Registration of the human actions is largely focused on group and individual spatial disposition and changing proximity to the screen (and thus the space occupied by the monkeys) and the nature of their movement, which can be interpreted as threatening (if abrupt) or inviting (if calm).

[77] Omnistereo only produces correct imagery for viewers located at the centre. However, inside the AVIE theatre the omnistereo images can be viewed comfortably from any position. This is the principal advantage of a cylindrical screen were any image distortion is continuous and therefore far less perceptible. This observation is based on the experiences of the many hundreds of visitors AVIE has already received (McGinity, M., Shaw, J., Kuchelmeister, V., Hardjono, A. & Del Favero, D. 'AVIE: a versatile multi-user stereo 360° interactive VR theatre' Proceedings *of the 2007 Workshop on Emerging Displays Technologies: Images and Beyond: The Future of Displays and Interaction*, San Diego, August 2007, vol. 252, New York: ACM (2007).

[78] Kenderdine, Shaw, Del Favero & Brown, 2007 op cit; Kenderdine 2007a op cit. 2007b op cit.

Figure 16. Scene from the Ramayana and the gathering of Monkeys at Kishkhinda. Source: British Library.

Figure 17. Young boy performing as Hanuman at Hampi. **Figure 18.** Monkeys at Hampi.

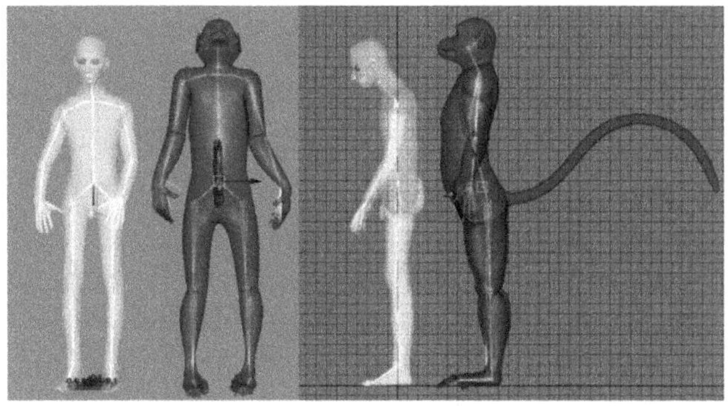

Figure 19. Rigging money models for animation.

Figure 20. Motion capture for Hampi-LIVE.

Figure 21. Graphic representation of Hampi-LIVE.

An interpretive matrix provides the mapping of various articulations of human behaviours to the scripted and improvisational range of monkey behaviours, and is time-sensitive enough to enable a 'narrative' development (and co-evolution). In Hampi-LIVE the mixed reality potentialities of direct real time interaction and co-evolutionary narrative development are thereby explored whereby sophisticated algorithms allow this tribe of monkeys to simulate and embody scenarios of internal and external social relations - internal to their own interests and external in their seemingly autonomous and spontaneous real time reactions and interactions with audience (fig. 21).

In scenario two, a virtual tourist who is wandering through a virtual Hampi scene with a video camera in her hand, suddenly notices that there are one or more real visitors in the AVIE space near her, and gestures to them to get them to move to one particular location in the room. Once she has succeeded in doing this, she takes her video camera, aims it at these visitors and seemingly films them for a while. When finished, this virtual tourist turns her virtual camera around and on its virtual viewfinder shows the visitors a replay of the film she has shot – which turns out to be a real movie of those visitors! What has actually transpired is that a real infrared camera mounted on the screen perimeter just above the virtual camera person's head has done this video recording, and by means of computer graphic techniques that recording has been composited onto the screen of the virtual camera's viewfinder.

For the visitor, there is an intense experience of surprise and engagement, as the computational dramaturgy has created a machine agent tourist who is seemingly able to gaze into, understand, and communicate with people in the real world, and even has possession of virtual apparatus that can record and play back this real world. This conjunction of the real and virtual provokes a hybrid dialogue on the notion of the 'touristic gaze'.[79] The hybrid and augmented nature of the experience is two-fold. Firstly, the visitors to Hampi-LIVE are made explicitly aware of the performance of a 'touristic gaze' - in this case through the virtual tourist and her video recording device - as they themselves are also surveying the same scene. Secondly, participants become objects in this scene as the virtual tourist turns her 'gaze' upon them and draws them into the record. The visitors become both visitors to the virtual site and objects within the virtual site.

Through the affordances of AVIE's innovative technologies, Hampi-LIVE is a modest experiment to transcend the common interpretive frameworks to become a site of mediation, an entanglement of people-things from the past and present occupation of the site. The theoretical connection between this work new narrative and concepts in post-processural archaeology is further explored below.

[79] Urry, J. *The tourist gaze*, 2nd edn (London: Sage Publications 2001).

5. Digital narrative, co-action and mediation

Concepts of digital narrative applied to new media remain predominantly uni-modal, and heritage researchers understand virtual heritage narrative as a derivative of conventional notions of virtual reality and cultural memory, whilst lacking an understanding of the complex, multi-dimensional quality of digital and cultural processes. Modelled on mimetic theory, many virtual heritage practitioners have theorized narrative in spatial terms as simulation[80] and recovery.[81] These uni-modal formulations flatten narrative into a one-dimensional, ready-made object, ignoring the multi-dimensional dynamics involved in a narrative generated through the interchange between ontologically divergent human and machine entities. Mediation, on the other hand, is a mode of engagement that takes us beyond narrative, for scholarly narrative obfuscates the multiplicity of material presence. Critically, mediation as it is envisaged by post-processual archaeologists such as Webmoor and Witmore, calls attention to the co-action of what has (by convention) been split apart - that is, subject and object - in accounts of representation. In this regard, mediation symmetrically shifts the "burden of knowing". The dichotomy moves away from a subject-versus-society pole - representing externalized, inert reality[82] - to find symmetry. In symmetrical archaeology, rather than nature and society poised across from each other on a horizontal axis, nature-society is seen as a complex *entanglement of people-things.*[83] Human beings are not (detached and singular intentional agents, but rather always are implicated in complex socio-technical assemblages."[84] Further, "Mediation (re)balances claims to know the world by excavating beneath representation as conventionally understood, and provides both an ontology of the co-creation of people-things and an epistemology not encumbered by the subject-world gap...."[85]

The framework advanced in the research addresses a need articulated by virtual heritage scholars to treat the heritage object as an evolving experience, or symmetrical experience (addressing arguments in hermeneutics, for example), in which the story told is not pre-rehearsed but emerges as an interactive dialogue between viewers and agents. Deleuze[86] and De Landa[87] theorized that

80 Baudrillard, J. *Impossible exchange*, Turner, C., trans. (London: Verso 2001).

[81] Sturken, M. 'Narratives of recovery: Repressed memory as cultural memory', in Bal, M. & Crewe, J. (eds), *Acts of memory: Cultural recall in the present,* Hanover, NH: University Press of New England, (1999): 231–48.

[82] Webmoor 2005 op cit.

[83] Webmoor, T. *Lessons from the real: Mediating people-things in a symmetrical archaeology* (2007) available online <http://traumwerk.stanford.edu:3455/Symmetry/743>, last accessed 30 June 2009.

[84] Webmoor, T 2005 op cit.

[85] Webmoor, T. & Witmore, C. 2005 op cit.

[86] Deleuze, G. *Cinema 2: The time-image*, Tomlinson, H. & Galeta, R. trans. (London: The Athlone Press 1989).

Deleuze, G. *Negotiations 1972–1990*, Joughin, M. trans. (New York: Columbia University Press 1995).

[87] De Landa, M. 'Virtual environments and the emergence of synthetic reason', in Dixon, J. & Cassidy, E. (eds), *Virtual futures: Cyberotics, technology and post-human pragmatism*, (London: Routledge, 1998),

narrative is a reciprocal process in which meaning is co-generated by intelligent agents and viewers as the result of a modest narrative of exploration. Narrative is a process that interweaves viewers and cinematic images in the production of new multi-layered events that simultaneously incorporate the past and present.[88] Providing autonomy to machine-agents balances the interactive initiative between virtual characters and viewers within virtual heritage. This equalizing of agent and participant transforms the encounter into an exciting and unpredictable drama in which events are co-produced. In this way, the research incorporates philosophical underpinning of symmetrical archaeology. The relative autonomy of the machine-agents is crucial in the Hampi-LIVE and distinguishes it from the majority of avatar-based works. It is a preliminary experiment in creating co-evolutionary narrative experiences in virtual heritage environments, underwritten by the potentiality of phenomenology and embodiment of place. As the machine-agents and participants learn to stimulate these scenarios through their actions, they co-evolve unique narrative experiences. Hampi-LIVE instigates a digital reformulation of the notions of Christopher Pinney' 'corpothetics' (corporeal, embodied aesthetics and somatic solidarity).[89] This reformulation is both convergent with and facilitated by the new technologies of immersion, presence and hybrid multimodal interaction.

Conclusion

At the intersections of culture, heritage and new technologies, and of recombinatory poetics, phenomenology and the visual, Zelinski observes that the media interface is both: "...poetry and techne capable of rendering accessible expressions of being in the world, oscillating between formalization and computation, and intuition and imagination."[90]

The works described in this paper hope to provide contexts for the imagination to flourish within the context of archeological landscapes. Post-processural archaeology provides researchers at the interface of new media and humanities with the theoretical foundation that inform methodological approaches to the interpretation of these landscapes. This paper has tried to demonstrate how these frameworks can applied, and provide foundation criteria for the future development of interactive and immersive digital heritage applications. This work proposes a shift from traditional approaches of 3D modelling in virtual heritage to a more somatic and kinaesthetic engagement and phenomenological encounter with place.

Acknowledgements

PLACE-Hampi and Hampi-LIVE are projects by Museum Victoria and iCinema Centre for Interactive Cinema Research, UNSW supported by an Australian Research Council Linkage Grant 2006-2010. The principal authors of the work are Sarah Kenderdine and Jeffrey Shaw. The Martin Gropius Bau evaluation survey of PLACE-Hampi was undertaken with ethics approval from UNSW, Australia. Ancient Hampi is an exhibition curated and designed by Sarah Kenderdine for the Immigration Museum and Museum Victoria, Melbourne. All images are copyright of the author unless otherwise stated.

65–78.

88 Kenderdine 2007b op cit.

89 See Kenderdine 2007a op cit.

90 Zelinski 2006, op cit., 277.

Bibliography

Bateson, G. & Bateson, M. C. *Angels fear: Towards an epistemology of the sacred* (Chicago: University of Chicago Press 1988).

Bateson, G. *Mind and nature: A necessary unity: advances in systems theory, complexity, and the human sciences* (New Jersey: Hampton Press 1979).

Bateson, G. *Steps to an ecology of mind: Collected essays in anthropology, psychiatry, evolution, and epistemology* (Chicago: University of Chicago Press 1972).

Baudrillard, J. *Impossible exchange*, Turner, C. trans. (London: Verso 2001).

Bennett, T. 'Civic seeing: museums and the organisation of vision', in S. MacDonald (ed.), *Companion to museum studies*, Oxford: Blackwell (2006): 263–81.

Dalsgaard, P. & Koefoed-Hansen, L. 'Performing perception—staging aesthetics of interaction', ACM *Transactions on Computer-Human Interaction* (TOCHI), Vol. 15, issue 3, Article No. 13: (2008): 1-33.

De Landa, M. 1998, 'Virtual environments and the emergence of synthetic reason', in Dixon, J. & Cassidy, E. (eds), Virtual futures: Cyberotics, technology and post-human pragmatism, London: Routledge, (1998): 65–78.

Deleuze, G. *Cinema 2: The time-image*, Tomlinson, H. & Galeta, R. trans. (London: The Athlone Press 1989).

Deleuze, G. *Negotiations 1972–1990*, Joughin, M. trans., (New York: Columbia University Press 1995).

Hodder, I. 1999, *The archaeological process: An introduction* (Oxford: Blackwell 1999).

Ihde, D. 'Material hermeneutics', in *Symmetrical archaeology*, Stanford: Theoretical Archaeology Group, (2005) available online <http://humanitieslab.stanford.edu/Symmetry/746>, last accessed 3 August 2007.

Kenderdine, S. 'Somatic Solidarity, Magical Realism and Animating Popular Gods: Place-Hampi 'where intensities are felt'', in Banissi, E., Burkhard, R. A., Grinstein, G., Cvek, U., Trutschl, M, Stuart, L., Wyeld, T. G., Andrienko, G., Dykes, J., Jern, M., Groth, D., & Ursyn, A. (eds) *Proceedings of the 11th European Information Visualization Conference*, IV07, Zurich, Switzerland, July, IEEE Comp Society, (2007a): 402-408.

Kenderdine, S. 'The Irreducible Ensemble: Place-Hampi', in (eds.), Kenderdine, S. & Wyeld, T. Docherty, M. 2007, (eds.), Proceedings of Virtual Systems and Multimedia 13th International Conference, VSMM 2007, Brisbane, Australia, September 23-26, Revised Selected Papers Series: Lecture Notes in Computer Science, Subseries:

Information Systems and Applications, incl. Internet/ Web, and HCI, vol. 4820, Berlin Heidelberg: Springer, (2007b): 58-67.

Kenderdine, S. & Schettino 'PLACE-Hampi: interactive cinema & new narratives of embodied cultural experience', *Inclusive Museum conference*, Istanbul June 2010. Abstract submitted.

Kenderdine, S., Shaw, J. & Kocsis, A., 'Dramaturgies of PLACE: Evaluation, Embodiment and Performance in PLACE-Hampi', Proceedings of the DIMEA/ACE Conference (5th Advances in Computer Entertainment Technology Conference & 3rd Digital Interactive Media Entertainment and Arts Conference) Athens, November (2009): forthcoming.

Kenderdine, S., Shaw, J., Del Favero, D. & Brown, E. 'Place-Hampi: Co-evolutionary narrative & augmented stereographic panoramas, Vijayanagara, India', in Kalay, Y., Kvan, T. & Affleck, J. (eds), *New heritage: New media and cultural heritage*, Abingdon: Routledge/ Taylor & Francis Books, (2007): 336–352.

McGinity, M., Shaw, J., Kuchelmeister, V., Hardjono, A. & Del Favero, D. 'AVIE: a versatile multi-user stereo 360° interactive VR theatre' Proceedings *of the 2007 Workshop on Emerging Displays Technologies: Images and Beyond: The Future of Displays and Interaction*, San Diego, August 2007, vol. 252, New York: ACM. (2007).

Olsen, B. 'Material culture after text: Re-membering things', in *Norwegian Archaeological Review*, vol. 36, no. 2, (2003): 87–104.

Pearson, M. & Shanks, M. *Theatre/Archaeology*, (London: Routledge, 2001).

Pinney, C. *Photos of the gods: The printed image and political struggle in India*, (London: Reaktion Books 2004).

Shanks, M. 'Archaeology and the visual' (2008), available online <http://documents.stanford.edu/ michaelshanks/144>, last accessed 30 June 2009.

Shanks, M. 'Media and archaeological futures', (2006) available online <http://documents.stanford.edu/ michaelshanks/113>, last accessed 19 May 2009.

Shanks, M. 'Symmetrical archaeology', in World Archaeology, vol. 39, no. 4, (2007): 589–96.

Shanks, M. 'Three rooms: Archaeology and performance', in Journal of Social Archaeology, vol. 4, no. 2, (2004): 147–80.

Shanks, M. *Experiencing archaeology* (London: Routledge 1992).

Sturken, M. 'Narratives of recovery: Repressed memory as cultural memory', in Bal, M. & Crewe, J. (eds), *Acts of memory: Cultural recall in the present,* Hanover, NH: University Press of New England, (1999): 231–48.

Tilley, C. *Body and image: Explorations in landscape phenomenology* (Walnut Creek: Left Coast Press 2008).

Tilley, C. *The materiality of stone: Explorations in landscape phenomenology* (Oxford: Berg 2004).

Tilley, C. *The phenomenology of landscape* (Oxford: Berg 1994).

Thomas, J. 'The great dark book. Archaeology, experience, and interpretation', in Bintliff, J. (ed.), *A companion to archaeology*, Oxford: Oxford University Press, (2004): 21–36.

Urry, J. *The tourist gaze*, 2nd edn (London: Sage Publications 2001).

Webmoor, T. 'Lessons from the real: Mediating people-things in a symmetrical archaeology', (2007) available online <http://traumwerk.stanford.edu:3455/ Symmetry/743>, last accessed 30 June 2009.

Webmoor, T. 'Mediational techniques and conceptual frameworks in archaeology', in *Journal of Social Archaeology*, vol. 5, no. 1, (2005): 54–86.

Webmoor, T. & Witmore, C. 'Symmetrical archaeology' (2005), available online <http://traumwerk.stanford. edu:3455/Symmetry/home>, last accessed 30 June 2009.

Witmore, C. L. 'Four archaeological engagements with place: Mediating bodily experience through peripatetic video', in *Visual Anthropology Review*, vol. 20, no. 2, (2004): 57–72.

Witmore, C. L. 'Vision, media, noise and the percolation of time: Symmetrical approaches to the mediation of the material world', in *Journal of Material Culture*, vol. 11, no. 3, (2006): 267–92.

The Fallacy of Reconstruction

Jeffrey T. Clark
North Dakota State University
Fargo, North Dakota, USA

Abstract

The idea of reconstructing the past has been part of archaeology almost since its inception as a discipline. Throughout most of the 20th century, the "reconstruction" of culture history, culture lifeways, or culture processes explicitly guided the work of most archaeologists. By the turn of the 21st century, the notion of "reconstructing" some ancient building or cultural setting had also become entrenched in the new domain of virtual archeology. It is the contention of this paper that the notion of "reconstructing" the past is not only a misnomer but one that has been detrimental to the discipline. This holds true for conventional archaeology as well as virtual. We should only be talking about "constructions" of the past and rarely, if ever, about reconstructions. We are *always* constructing models, whether visual, verbal, or some other type, which are tools for understanding, not statements of reality. The criticisms that have often been leveled against virtual visualizations have been largely due to calling, and thinking about, visualizations as reconstructions of some aspect of the past rather than regarding them for what they are—models. Calling these models reconstructions is not only fallacious but has hindered the acceptance and use of virtual archaeology by the larger professional community.

Key words: *reconstruction, simulation, model, interpretation, multivocality, virtual archaeology*

1 Introduction

In one of Shakespeare's most memorable lines, Juliet emotes to Romeo:

"What's in a name? That which we call a rose By any other name would smell as sweet" (*Romeo and Juliet, II, ii, 1-2*).

The point of her statement is that one's family name (Capulet or Montague) is but a convention and not of true significance. The fact is, though, that Romeo and Juliet are plagued by their family names, which carry with them an enormous weight of meaning, ultimately leading to tragedy. In the end, they cannot escape who they are even by vowing to reject their names and the baggage (families and feud) they carry.

This is an odd beginning, perhaps, to a paper on archaeology, but it points to the central theme to be discussed in this paper. That theme can be expressed in two points that at first glance might appear a bit contradictory, although I hope to show that they are not: 1) no matter what name or term one gives to something, the name does not change the essential character of that thing, and 2) terminology, or the names or words we call things, does matter.

To be more specific to the point of this paper, archaeologists may say they have created a "reconstruction" of some facet of the past, but in fact they have not, and with few exceptions cannot, "reconstruct" the past; *one can only construct models or simulations of the past*. To claim to have done so is a mistake. At the same time, by calling one's *con*struction a *re*construction, one misleads the viewer of that model, both professional and lay, and sometimes even oneself. And that is detrimental to what archaeologists are trying to accomplish in both research and education. This is so whether one is doing traditional archaeology or cyber-archaeology.

2 The Power in Words

There is indeed power in words. Words, including and perhaps especially names and terminology, *reflect* as well as *shape* what we think about things. Those terms convey meaning to others whose perceptions are then influenced, subtly or profoundly. This has long been well understood by wordsmiths, and seen most effectively in politics and the military. The number of examples that could be given is large, indeed, but I provide just two. In the 1980s, the term "misspoke" was used repeatedly by the administration of President Ronald Reagan in clarifications of what the President had said rather than simply acknowledging that the president had made a mistake. This application of the term to presidential politics was new. But in order to stem the impression that Reagan made mistakes more frequently than one would want of a President, it was couched as if he really knew the correct information, he just didn't say it correctly. It should be noted that that application of the term "misspeak" is not fully consistent with the definition, but the re-characterization to deflect criticism was, for the most part, successful. For a more current example, the administration of George W. Bush engaged in "enhanced interrogation techniques," not torture. The more a term or name is used, the more it influences how people perceive and think about things—say it enough and people will start to believe it.

To say that one has "reconstructed" something is to infer that one has re-created it just as it was. That would mean that one *knows* what it used to be and has replicated it, which implies a degree of "truth." One may be able to reconstruct a broken pot from the collected pieces, or a house that burned down by following the original blueprint and photographs, but one cannot *re*construct a house that existed at some point in the past and for which we now have fragmentary information from feature traces and artifacts. What one *can* do is construct a model, verbal or visual, of how that house may have looked, given the available evidence. The more complex the phenomenon that we are examining archaeologically, the more difficult it is to "reconstruct."

To say that one has actually created an archaeological reconstruction is in most cases fallacious. This is not to say that it is malicious, but simply a legacy of the discipline that has been perpetuated. Moreover, as developments in computer technology have made it possible to create evermore realistic virtual visualizations of past places (buildings and landscapes) and peoples, the creators have taken to calling them "reconstructions." This in turn has brought criticism of the use of, and in some cases the general practice of, virtual modeling. The more detailed the visual models, the more they approach photo-realism, the more likely they are to be criticized, at least by some archaeologists. These criticisms, however, are in large part a consequence of inappropriate terminology and the related misconception of what is actually being produced.

3 Archaeological Reconstruction as Legacy

In this section I will provide an admittedly sketchy historical summary of archaeology as it relates to the notion of reconstruction. This is not intended to be a thorough coverage; the complexity of the historical development of archaeology is well beyond the scope and intent of this article. Instead, I wish to establish the legacy of the concept of reconstruction, which is a big part of why the idea is still with us despite major shifts of the intellectual winds in archaeology. As a side note, I would add that I use the term legacy in both its senses: of an idea handed down from the past, and, as used with computer terminology, as old and outmoded.

The notion of "reconstructing the past" has long been part of archaeology as an academic discipline. In the early decades of the 20th century, American anthropology forged its own character, its own perspective and tradition. Three specific conditions were critical factors in the making of American anthropology: 1) there were extant Native American peoples whose cultures, languages, and physical attributes were being studied by anthropologists; 2) relatively recent ethnohistoric documents could be used to establish an "historic baseline" for individual Native American cultures; and 3) traces of past societies seemed naturally to meld together with the ethnographic observations and ethnohistoric documentation. As a result, the distinctive four-field perspective came to characterize American anthropology, and archaeology drew heavily from cultural anthropology while also informing it. In the first few decades of the 20th century, what Waldo Wedel[1] eventually dubbed the Direct Historical Approach came to be widely employed for understanding prehistory in North America, as archaeologists used historic baselines and analogy derived from ethnography and ethnohistory to work back in time, using the known to reconstruct the unknown .[2]

Franz Boas and his many students had a profound influence on the development of anthropology in the first half of the 20th century. The perspective that came to be known as historical particularism emerged by which one comes to understand a given culture by studying the specifics of its particular historical development. A central task of anthropology, at least at that stage in its development as a discipline, was to collect copious and detailed information on all facets of a culture. The link of archaeology to anthropology translated that task to the search for detailed material data, or trait lists, on past cultures. Such trait lists for different points in time at the same location essentially came to be regarded as the reconstruction of the culture history of those past peoples.

Boas's most influential student was Alfred Kroeber, who played an important role in bringing archaeology into anthropology. Although he did not consider himself an archaeologist, he wrote about, taught, and did field work in archaeology. He regarded human history as critically important for understanding modern human societies, and he was instrumental in moving archaeology away from a staunch focus on classification of prehistoric monuments, sites, and artifacts that characterized the 19th century approach to a more anthropological concern with the cultures that produced those remains. Indeed, Kroeber helped re-focus archaeology on efforts to reconstruct culture histories. In 1937, describing the term archaeology for the *Encyclopedia of the Social Sciences*, he wrote: "Archaeology, by etymology the study of beginnings, has historical reconstruction for its objective"[3]

The notion that *culture historical reconstruction* was the primary goal of archaeology became entrenched in the discipline. With the increasing collection of data and the compilation of trait lists, there came a perceived need for the classification of cultures into assorted divisions and subdivisions (e.g., the Midwest Taxonomic System, The Southwest Pecos Classification, Central California Classification, and others). In association with this came the normative view of culture in which cultural traits are the product of shared cultural beliefs and norms of behavior. The norms guided individuals in the manufacturing of artifacts and structures, their use, and even their disposition. Thus, the researcher could uncover the cultural norms by study of the material culture, and from that "reconstruct" the culture over time.

[1] Waldo Wedel, *The Direct-Historical Approach in Pawnee Archeology* (Washington, D.C.: The Smithsonian Institution, 1938).

[2] Julian H. Steward, "The direct historical approach to archaeology," *American Antiquity* 7 (1942).

[3] Alfred. L. Kroeber, "Archaeology." In, *Encyclopedia of the Social Sciences, Vol. 2.*
(1937): 163).

Beginning around 1940, a new aspect of reconstruction came to the fore and gained considerable strength through the 1950s. That was the *reconstruction of past lifeways*. The key to lifeways was taking a broader approach to the study of prehistory by more effectively using ethnographic analogy, by looking at all aspects of life in the past, not just the artifacts and structures, and by bringing environmental context clearly into the mix. The concern with problems of cultural process in anthropology can also be found in the early 1940s with the studies of lifeways by scholars such as Steward, Bennett, Childe, Caldwell, and others who, over the next decade or more, looked to explore the adaptive patterns of cultures in relation to their environments. Perhaps the strongest influences in archaeology were the works of J. Grahame Clark at Star Carr, England[4] and Gordon Willey at Virú Valley, Peru.[5] An interest in the processes of cultural adaptation was by its very nature an interest in and study of the processes of change.

The emergence of reconstruction of past lifeways as a goal of archaeology took place in conjunction with the increasingly held view that archaeology was not sufficiently anthropological. That was a central point in Walter Taylor's 1948 classic work, *A Study of Archeology*, in which he explicitly called for what he called a "conjunctive approach." Such an approach would turn away from normative, historical, descriptive archaeology that dominated the first half of the 20th century in favor of a holistic, culture-based approach. Ten years later, another classic work, *Method and Theory in American Archaeology* by Gordon Willey and Philip Phillips, sought to emphasize the connection of archaeology to anthropology. While that work was still strongly oriented towards description, classification, and reconstruction, it lamented, as Lewis Binford[6] pointed out, the lack of an explanatory level in archaeology and also suggested the value of what they called "processual interpretation."[7] They defined "processual interpretation" as "the study of the nature of what is vaguely referred to as the culture-historical process."[8] Their phrase, "archaeology is anthropology or it is nothing" was picked up a few years later by Binford[9] who championed it, as he and others also championed the idea of Willey and Phillips[10] that "we are no longer asking merely what but also how and even why."

It is well known, of course, that in the 1960s and 1970s, discontent with the conventional pursuit of the reconstruction of culture history as the driving purpose and end point of archaeological research, in the context of a tumultuous socio-cultural milieu in which authority was challenged and science was championed, resulted in the emergence of a new direction in archaeology. The influential writings of Binford, Flannery, Spaulding, Deetz, Watson, and many others expressed and fueled this discontent, leading to what became known as the New Archaeology (or as it came to be written, New Archeology). The new archaeology was actually a rather diverse set of approaches that seemed to coalesce around an emphasis on the scientific approach to explaining the process of cultural development and change—that is, on the how and why—over culture history as the goal of research. Philosopher of science Carl Hempel's[11] deductive-nomological (for scientific explanation) and hypothetico-deductive (for scientific confirmation) paradigms were taken by many to be the only acceptable approach for a scientific archaeology.[12] Eventually, the term "processual archaeology" came to be applied to this school of thought, emphasizing the focus on explaining cultural processes and cultural change. Processualism became the dominant, though not exclusive, approach in American anthropological archaeology.

Despite the dominance of the processual approach, the old notion that the ultimate goal of archaeology is some form of reconstruction was largely retained. In an extremely influential work, Binford identified three aims of archaeology: 1) the reconstruction of culture history, 2) the reconstruction of past lifeways, and 3) the study of cultural process.[13] The idea of *reconstructing* cultural process was also advocated by some. Writing in 1982, American historian Thomas Schlereth characterized archaeology as composed of "*static* reconstructionists and *process* reconstructionists."[14] The former group he essentially equated with those interested in the reconstruction of cultural history and past lifeways, and the latter with the processualists.

Ironically, what began with a concern to be more mindful of culture in its totality led ultimately to an approach in which cultures, as unique, historical forces in peoples' lives, were effectively downplayed. Behavior rather than culture had for many become the focus of archaeological analysis; processual forces and behavioral adaptations were the key to explaining the past.[15] Even key players in the new archaeology movement came to realize that the wheels were coming off and some correction was needed. Culture needed to be brought back into the

[4] J. Grahame D. Clark, *Star Carr* (Cambridge: Cambridge University Press, 1954).

[5] Gordon R. Willey, *Prehistoric Settlement Patterns in the Virú Valley, Peru* (Washington D.C.: Smithsonian Institution, Bureau of American Ethnology, 1953).

[6] Lewis R. Binford, "Archaeological Perspectives." In, *New Perspectives in Archeology*, (S. Binford and L. R. Binford, eds., Chicago: Aldine Publishing CO, 1968), 7.

[7] Gordon R. Willey and Philip Phillips, *Method and Theory in American Archaeology* (Chicago: University of Chicago Press, 1958), 5.

[8] *Ibid*, 5-6.

[9] Lewis R. Binford, "Archaeology as anthropology" *American Antiquity* 28 (1962).

[10] Willey and Phillips, 6.

[11] Carl Hempel, *Philosophy of Natural Science* (Englewood Cliffs, NJ: Prentice-Hall, Inc., 1966).

[12] Patty Jo Watson, Steven A. Le Blanc, and Charles Redman, *Archaeological Explanation: The Scientific Method in Archaeology* (New York: Columbia University Press, 1984).

[13] Binford., 1968.

[14] Thomas J. Schlereth, "Material Culture Studies in America, 1876-1976." In, *Material Culture Studies in America: An Anthology* (T. J. Schlereth, ed., pp. 1-78. Walnut Creek, CA: AltaMira Press, 1982), 46 (emphasis in original).

[15] Michael B. Schiffer, *Behavioral Archeology* (New York: Academic Press, 1976).

picture[16], and the claims of pure scientific objectivity were acknowledged as incorrect.[17]

Just as dissatisfaction with traditional archaeology led to the processual movement, by the 1980s dissatisfaction with the shortcomings of processual approaches (particularly the emphasis on objective science and the deductive-nomological intellectual straightjacket) led to a non-processual movement, of sorts, inspired by the writings of Hodder, Shanks, Tilley, and others. While not a movement in the sense of a unified perspective, they were unified in the sense of what they were not, which was new archaeology processualists or behavioralists. The broad umbrella term "post-processual" came to be applied to these varied approaches, although Renfrew[18] perhaps more appropriately dubbed them "anti-processual." The varied approaches under the post-processual umbrella include interpretive archaeology, critical archaeology, structuralist, feminist, hermeneutic, cognitive, symbolic, postmodern, and other approaches. Processual critics charged, among other things, that post-processualists were too long on theory and too short on methodology and practical (in the field) substance, and that the interpretive relativism of post-processualists meant that any interpretation is as good as the next, a position which they believed was untenable. Many post-processualists have responded that there is not, and should not be, a single methodology any more than there should be a single interpretation or voice on the past, but that does not mean that all interpretations are equally valid.

As highlighted by Hodder[19], post-processualism was a "reactionary" approach—reactionary in the sense of a reaction to processualism, especially in its extreme formulations, and in its return to an appreciation of culture as historically relative (as it was characterized in the writings of many who were criticized by the new archaeologists) and meaningfully constituted (as characterized by cultural anthropologists). Hodder[20] has argued that by the early 1990s one could speak of a post-processual era more than a post-processual movement in archaeology because of its lack of a unifying thrust beyond the attempt to move archaeology away from the strict focus on scientific methodology and the elucidation of cross-cultural laws.

While the contentions of post-processual thinking are diverse, I will focus on what has become known as *interpretive archaeology,* which is arguably the most influential approach, as well as the most relevant to this discussion. From this perspective, interpretation of archaeological evidence is inescapable and essential, and context is critical for interpretation.[21] Archaeologists engage, in an interpretive discourse, a hermeneutic reading, of observations rather than an unbiased, detached, socially sterile discovery. Facts are, it is claimed, theory-laden and subjective. Evidence is contextual and contingent. By context is meant more than the archaeological spatial and temporal context, but includes the cultural and historical contexts as well as the contemporary context of the interpreter. Actions in the past occurred and had meaning within a cultural context, and the archaeological observation of the traces of those actions—which is to say the specific material traces—take place within the cultural context and understanding of the observer. Evidence is always viewed from the perspective of the prejudices, preconceptions, and predilections of the observer. Objectivity, scientific or otherwise, is simply not possible. From a generalized post-processual viewpoint, the idea of "the" reconstruction of the past at any site is inconsistent with notions of interpretation, context, multivocality, and agency.

Despite the emergence and importance of the assorted post-processual approaches, I contend that most archaeologists in the United States, especially those who are active field archaeologists, still largely operate with an essentially "processual" mindset, minus the explicit search for cultural laws and dogmatic adherence to a deductive nomological approach. Thus, the view that archaeology is still about reconstructing the past has remained strong, if not pervasive.

4 The Fallacy of Reconstruction

It is the contention of this paper that the emphasis on reconstruction as central to conventional old and new archaeology has been detrimental to the discipline. It is detrimental because it conveys a false sense of knowledge. This in turn makes it unduly difficult for new interpretations that stand in contrast to the conventional reconstruction to get a fair hearing, thereby impeding advances in our understanding.

The notion that the term and concept of "reconstruction" is fallacious and should be abandoned is hardly new. Indeed, Walter Taylor, long recognized as an inspiration for the new archaeology, made the point clearly in his 1948 seminal work. Though that work was often cited by new archaeologists, his argument on the fallacy of reconstruction was essentially ignored. I will cite Taylor at length on this because he makes the point so well.

[16] Kent V. Flannery, "The golden Marshalltown: a parable for the archaeology of the 1980s," *American Anthropologist* 84 (1982).

[17] Lewis R. Binford and Lewis R. Sabloff, "Paradigms, systematics and archaeology," *Journal of Anthropological Research* 38 (1982).

[18] Colin Renfrew, "Towards a cognitive archaeology." In, *The Ancient Mind. Elements of Cognitive Archaeology* (C. Renfrew and E. B. W. Zubrow, eds., pp. 3-12. Cambridge: Cambridge University Press, 1994).

[19] Ian Hodder, "Theoretical archaeology: a reactionary view." In, *Symbolic and Structural Archaeology* (I. Hodder, ed., pp. 1-16. Cambridge: Cambridge University Press, 1982).

[20] Ian Hodder, "Interpretive Archaeology and its role" *American Antiquity* 56 (1991).

[21] E.g., Ian Hodder, "Theoretical archaeology: a reactionary view." In, *Symbolic and Structural Archaeology* (I. Hodder, ed., pp. 1-16. Cambridge: Cambridge University Press, 1982); *Reading the Past: Current Approaches to Interpretation in Archaeology* (Cambridge: Cambridge University Press, 1986); Ian Hodder, et al., (eds.), *Interpreting Archaeology. Finding Meaning in the Past,* (London: Routledge, 1995); Michael Shanks and Christopher Tilley, *Re-constructing Archaeology* (Cambridge: Cambridge University Press, 1987); Julian Thomas, *Interpretive Archaeology. A Reader* (Continuum International Publishing Group, 2001); Christopher Tilley (ed.), *Interpretative Archaeology* (Oxford: Berg, 1993).

This point was so important to Taylor that in his 1972 article in the important new archaeology volume edited by Mark Leone, he cited this passage from his 1948 book and notes somewhat despairingly that the new archaeologists were setting themselves up for failure by ignoring cultural context and espousing an impossible goal of "reconstructing past lifeways".[22]

> ...it will be well to comment on the fallacy and mental hazard engendered by the use of the term *historical reconstruction*. The words *re*construction and *re*synthesis are fundamentally erroneous and have been responsible for much loss of confidence, particularly among the anthropologists.... If it cannot be told for sure whether past actuality has, or has not, been recreated in detail or in essence, it cannot be claimed that these contexts are, or are not, *re*syntheses or *re*constructions. These terms imply a re-building to exact former specifications which, from the above, are not verifiable and, hence, not knowable. The unknowable cannot be taken as a standard value. Therefore, arguments both for and against historical reconstruction in anthropology or in any other [historical] discipline are irrelevant, and it becomes apparent that the work of all historical disciplines really leads to construction and synthesis, not reconstruction and resynthesis. From this, it is further apparent that the real task of the students in historical disciplines settles down to seeing how sound, how plausible, and how acceptable their constructions can be made. Neither the anthropologists nor the historians should use the term *reconstruction* and thus make himself feel inadequate because he knows that his research will never permit him actually to reconstruct the life of past times with certainty and completeness. Rather, he should realize that even the contexts written from the best and fullest archives are constructions and that the differences lie in the nature of the respective data, not in the procedures or basic theoretical factors.[23]

Taylor's argument, in both instances, fell on deaf ears. Despite his clear and forceful statement, the notion of reconstruction remained in archaeology. So if not reconstructions, what do we call this process and product? We construct *models*, and whether one chooses to call them scientific models or interpretive models, the essential features are the same. There are, of course different types of models—verbal, mathematical, graphical, visual, simulation, and more. But they are always models, and only models. With the idea that we are dealing with models, I will briefly review what I see as the critical qualities of models in general.

[22] Walter W. Taylor, "Old wine and new skins." In, *Contemporary Archaeology: A Guide to Theory and Contributions* (M. P. Leone, ed., pp. 28-33. Carbondale: Southern Illinois University Press, 1972).

[23] Walter W. Taylor, *A Study of Archeology* (Memoir No. 69. *American Anthropologist* 50(3), Part 2, 1948), 35-6.

5 Models

I have long agreed with Taylor on the fallacy of reconstruction. In 1978, John Terrell and I published an article that reviewed the state of archaeology in Oceania.[24] In that piece we criticized the traditional goal and practice of culture historical reconstruction, which we characterized as the writing of "scenarios" that were little more than just-so-stories. Such accounts do not disentangle fact from fancy, and despite the fact that much was still unknown, the scenario was typically presented as if it was true—after all, it was the "reconstruction" of the past. In place of the culture historical scenario or reconstruction, we advocated an experimentalist approach based on the method of multiple working models (adapted from Chamberlain[25] in geology and others who argued for a method of multiple working hypotheses). While our focus in that article was Oceania, the critique was intended to be taken as applicable to the practice of archaeology in any region of the world.

The role of models and the modeling process was recognized by many in the new archaeology school. David Clarke and colleagues discussed modeling thoroughly[26], and many archaeologists, especially those drawing from other disciplines, employed various modeling strategies and methods, such as systems theory modeling, mathematical modeling, geographic modeling, graphical modeling, and other approaches. Post-processual archaeologists also accepted the use of models as potentially useful fictions rather than as explicitly scientific tools.[27] But somehow along the way, the notion of model became distorted. A model was something that one may or may not use on the way to the ultimate goal of reconstruction. Too many people lost track of the fact that a reconstruction is just another model, not a statement of discovered truth.

Models are statements of relationships between multiple variables that provide a description and/or explanation of some phenomenon. Models are *not* the endpoint of the research, though they are too often regarded as such, especially when they are tagged with the term "reconstruction." They are tools that are constructed in order to help us think about and better understand some complex phenomenon. Models are produced from the interplay between empirical observations, assumptions, logical interpretation and extrapolation, and creative imagination. As Gibbon has noted, when archaeologists are properly engaged in the modeling process, "no commitment needs to be made to the view that *this* is what necessarily happened in the past" (emphasis in original).[28]

[24] Jeffrey T. Clark and John Terrell, "Archaeology in Oceania," *Annual Review of Anthropology* 7 (1978).

[25] Thomas C. Chamberlain, "The method of multiple working hypotheses," *Science* 15 (1890).

[26] David Clarke (ed.), *Models in Archaeology* (London: Methuen, 1972).

[27] Shanks and Tilley, 1987.

[28] Guy Gibbon, *Anthropolgical Archaeology* (New York: Columbia University Press, 1984), 47.

By definition, models are not the real thing; they are simplifications. As simplifications, something is left out, and the models are thereby always false. Information is left out because it is unknown or because it would provide unnecessary detail given the purpose of the model. Models are perhaps most useful when one is faced with incomplete information, which is always the case in archaeology. In a sense, then, by constructing a model one may be *selectively adding information*. One is likely to be adding more than what is known from direct observation in order to arrive at some reasonably coherent representation. From an interpretive archaeology perspective, one is always adding information based on the inescapable subjectivity, the "situatedness," of the observer. The selective addition of information is the case whether one is dealing with a verbal model or visual model, although, as discussed below, some see it is an issue of particular concern with visual models. The selection of information is based upon what one knows, thinks (consciously and subconsciously), assumes, and surmises.

While all models are abstractions and to varying degrees false, it is too often not sufficiently appreciated that different models can also be equally indeterminately "true," or at least useful. The more complex the phenomenon to be modeled, the more likely it is that there will be multiple models that are equally valid in the sense that each may provide a reasonable representation of that phenomenon. Writing on the Classic Maya collapse, Dimitri Shimkin stated: "Above all, it must be stressed that the acceptability of a model does not disprove the possibility of an indefinite series of other models, excepting only for its exact negative."[29] As phrased by Clark and Terrell: "For any set of archaeological data bearing on the complexity of the world, past or present, invariably there is more than one construction or model that could be built to account for the evidence known."[30]

A model is but a limited perspective on, and interpretation of, something too complex and/or too incomplete, to be known in its entirety. As an analogy, consider a ball of clay composed of swirling, multi-colored bands. If one were to take a slice through that ball in multiple places, the face of each slice would look different from the others. Each slice could be taken as a model of the whole, in which case each would be simultaneously equally true—an accurate description of the slice—and false—because it describes only a particular slice or perspective of the whole. A model is simply one slice through a complex ball of uncertain composition.

At the same time, when dealing with archaeological models, not all perspectives are as useful as others. It may well be that some models provide better fits with empirical observations than others. But as tools, the proper evaluation of a model depends on how useful it is for the purpose for which it is intended. While some models are for research, others are for education (at different levels), and others are for general public satisfaction and enlightenment. Furthermore, even models that do not provide equally good fits with observation or testing results may be informative for research purposes. Through the modeling process, and by testing or exploring ideas derived from the models, we hope to learn something we did know before.

Every presentation of the past by archaeologists, whether textual, graphic, or virtual, is the presentation of a "selected" set of information that constitutes an *interpretation*. It is an interpretation of what is important and what is not, of what the physical traces left to archaeologists might mean, and of how the whole comes together to constitute some sort of coherent pattern of meaning. Along with the social, political, and historical contexts of the interpreter that exert influences, there are the individual's presumptions, preconceptions, and predilections. The upshot is that for any complex modeling application, the best approach is to develop multiple working models, each reflecting differing assumptions and possible interpretations. We then do not fixate on finding *the* model that represents—or reconstructs—the truth. In essence, a method of multiple working models is a scientific expression of multivocality. It embraces the skepticism of producing *the* truth (for whatever reason one wishes to proffer) that is central to interpretive archaeology, while also providing an approach to scientific testing of hypotheses that one can derive from the models. In a sense, therefore, the method of multiple working models bridges the intellectual gap between post-processual and processual approaches.

6 Virtual Archaeology

The idea that archaeologists deal with reconstructions has carried into cyber-archaeology. In fact, the use of the term reconstruction is probably now more widely used in virtual archaeology than in conventional archaeology. It is commonplace at professional conferences in which computer applications are prominent (e.g., CAA, VSMM, VAST, and others) to see publications and papers titled and focused on a "reconstruction" of this or that building or site. That use is most often as fallacious as the textual reconstructions against which Taylor railed some 60 years ago; perhaps even more so because of the persuasiveness of the medium (seeing is believing). Consequently, it is important to examine the extension of the fallacy of reconstruction into this most modern approach in archaeology.

In the 1980s, a small group of archaeologists began exploring the application of computer graphics to archaeology for the creation of visualizations of archaeological remains. The use of virtual reality technology, however, was constrained by the limitations of computing power and software, thus limiting the

[29] Dimitri Shimkin, "Models for the Downfall: some ecological and culture-historical considerations." In, *The Classic Maya Collapse* (T. P. Culbert, ed., pp 269-99. Albuquerque: University of New Mexico Press, 1973).

[30] Clark and Terrell, 1978, 300.

reality component of virtual reality. The use of computer graphics for visualizing past structures and artifacts came to be called "virtual archaeology" by many. The concept of virtual archaeology was somewhat formalized in a classic paper by Paul Reilly in 1990 in which he defined virtual archaeology.

> The key concept is virtual, an allusion to a model, a replica, the notion that something can act as a surrogate or replacement for an original. In other words, it refers to a description of an archaeological formation or a simulated archaeological formation.[31]

The graphic modeling of virtual archaeology improved markedly over the next decade, with advances in the hardware and software capabilities, along with improving skills of the modelers, making it possible to create evermore realistic visualizations of past places (buildings and landscapes) and peoples. Correspondingly, the creators of these visualizations seemed to lose sight of Reilly's conception of virtual formulations as models, as simulations, and took to calling them "reconstructions" in the same sense as a culture historical reconstruction, which as noted above was simply in keeping with long-standing practice in archaeology. The use of the term reconstruction, however, is as misleading and fallacious for virtual archaeology as it is for conventional. For models in virtual archaeology, I will use the term *virtual simulations* to emphasize that they are digital, visual, immersive, and experiential rather than computational (e.g., agent-based models, predictive models, etc.). But they are always models, and only models, with all of the benefits and limitations of models.

As the so-called "virtual reconstructions" became more realistic, they engendered criticisms of the use of, and in some cases the general practice of, virtual reality modeling in archaeology. By the end of the 1990s, a debate was in full swing, and continues today, regarding the benefits and drawbacks of "virtual site reconstructions".[32] It is tempting to summarize the benefits of virtual simulation, which are notable, but that has been done by others and would deviate from the central point.[33] Instead, I will comment on the three most often cited criticisms that I think relate to the misconstrual of virtual simulation as virtual reconstruction.

The most often cited problem is the potential for visualizations to foster too great a sense of reality; to impart a sense of "truth" that in fact does not exist, or at the very least has not been documented.[34] This is especially so when the virtual simulation is done with a high level of realism, even approaching photo-realism. This criticism strikes me as comparable to saying that one should not create too compelling or complete a verbal or other model of some archaeological phenomenon because it might be too convincing to the readers who may take it to be more factual than it really is. Terminology is a critical factor. If one calls the model a reconstruction, whether virtual or otherwise, that false sense of knowledge may well be the consequence, intended or unintended. The terminology, in conjunction with the visual completeness, is what leads to criticism such as that by Shanks who opines that the "dream" of virtual archaeology is "that eventually, with so much data at hand, it will be possible for archaeological science to fill in the gaps," reflecting "an impossible archaeological desire to bring back the dead."[35] Shanks admits to creating a bit of a caricature with his comments, but as I have tried to show in this paper, the basis of that caricature can be understood when taken in the historical context of claims at reconstruction. I maintain that the criticism is, at the very least, blunted if one calls the visualization a simulation or model based on the interpretation of available data rather than a reconstruction of the past.

A related claim is that, because of the extraordinary power of the visual image to engender belief in its accuracy and authenticity, modelers should provide some mechanisms for allowing the viewer to know the *actual* level of accuracy and authenticity of the various components of the virtual representation. Without such mechanisms, virtual reconstructions cannot be regarded as valid archaeological

[31] Paul Reilly, "Towards a virtual archaeology." In, *Computer Applications in Archaeology 1990* (K. Lockyear and S. Rahtz, eds., pp. 133-39. Oxford: British Archaeological Reports, 1990), 133.

[32] E.g., Juan A. Barceló, "Visualizing what might be: An introduction to virtual reality techniques in archaeology." In, *Virtual Reality in Archaeology* (J. A. Barceló, M. Forte, and D. H. Sanders, eds., pp. 9-35. Oxford: BAR, 2000); Maurizio Forte, "About virtual archaeology: Disorders, cognitive interactions and virtuality." In, *Virtual Reality in Archaeology* (J. A. Barceló, M. Forte, and D. H. Sanders, eds., pp. 247-59. Oxford: BAR, 2000); Harrison Eiteljorg, II, "The Compelling Computer Image - a double-edged sword," *Internet Archaeology,* Issue 8 (2000); Stella Sylaiou and Petros Patias, "Virtual reconstructions in archaeology and some issues for consideration," *Imeros* vol 4 (2004); Graeme P. Earl, "At the edges of the lens: Photography, graphical constructions and cinematography." In, *Digital Archaeology. Bridging Method and Theory* (T. L. Evans and P. Daly, eds., pp. 191-209. London: Routledge, 2006); Donald. H. Sanders, "Why Do Virtual heritage?" In, *Digital Discovery: Exploring New Frontiers in Human Heritage. CAA 2006. Computer Applications and Quantitative Methods in Archaeology. Proceedings of the 34th Conference, Fargo, United States, April 2006* (J. T. Clark and E. M. Hagemeister, eds., pp. 427-36. Budapest, Hungary: Archaeolingua, 2007).

[33] E.g., see: Maurizio Forte and Alberto Siliotti, eds., *Virtual archaeology: Re-creating Ancient Worlds* (New York: Harry N. Abrams, Inc., Publishers, 1996); Donald. H. Sanders, "Archaeological publications using virtual reality: case studies and caveats." In, *Virtual Reality in Archaeology* (J. A. Barceló, M. Forte, and D. H. Sanders, eds., pp. 37-52. Oxford: BAR, 2000); Donald. H. Sanders, "Why Do Virtual heritage?" In, *Digital Discovery: Exploring New Frontiers in Human Heritage. CAA 2006. Computer Applications and Quantitative Methods in Archaeology. Proceedings of the 34th Conference, Fargo, United States, April 2006* (J. T. Clark and E. M. Hagemeister, eds., pp. 427-36. Budapest, Hungary: Archaeolingua, 2007).

[34] E.g., P. Miller and Julian Richards, "The good, the bad, and the downright misleading: archaeological adoption of computer visualization." In, *Computer Applications and Quantitative Methods in Archaeology 1994* (J. Huggett and N. Ryan, eds., pp.19-22. Oxford: BAR, 1995; Nick S. Ryan, "Computer based visualisation of the past: technical 'realism' and historical credibility." In, *Imaging the Past: Electronic Imaging and Computer Graphics in Museums and Archaeology* (T. Higgins, P. Main, and J. Lang, eds., pp. 95-108. London: British Museum Press, Occasional Paper 144), 1996); and conversations at every conference on virtual archaeology.

[35] Michael Shanks, "Listening to things: augmented reality and the science question in archaeology. The Science Question in Archaeology" http://documents.stanford.edu/michaelshanks/77 (2006).

expressions. The question is, to what extent, and by what means do we make uncertainty explicit? Various methods are being used by some modelers today to try and provide some level of information on accuracy and authenticity. These include use of color schemes, color intensity, image clarity, degrees of transparency, ghosting, labeling, hyperlinks to explanation pages, and scales by which the creators indicate their level of confidence in the accuracy of an element.[36]

But, all of these methods still come down to interpretation and judgment of confidence or certainty, which are still open to question if one is purporting to be reconstructing the past (as it was) rather than constructing a model reflecting an interpretation of what some specific time and place in the past may have been like. The fact is, one might be very confident, but quite wrong—a not uncommon occurrence in archaeology. One could argue, therefore, that specifying a high level of confidence is in fact *more* misleading than simply being explicit that one has created a model that reflects one interpretation. Because there is a multimodality and multivocality to interpretation, it will always be difficult to arrive at a consensus of confidence in any simulation.

The fact is, however, that most virtual simulations do not employ any mechanism by which the level of accuracy is provided, so this issue remains a point of debate. Certainly it is advisable to be able to provide access to a list of sources that influenced the creation of the virtual simulation, just as one would with a conventional archaeological model. But is it really more incumbent to do so for virtual simulations than any other model? To reveal every piece of evidence that factored into every element of any complex model might be too much for anyone to do. Moreover, as Earl notes, "to show all of the conflicting information available is an impossible project."[37] To force that on the creators of virtual simulations would have a chilling effect. Virtual simulations are produced for research, education, entertainment, or some combination of these (e.g., edutainment), and whether, in what way, and to what extent one provides information on sources for and confidence in a given simulation will depend on the purpose of the simulation.

A third issue of concern is the potential vagary as to what is being modeled in any given simulation. As we all know, archaeological sites typically are not instances in time but, instead, represent spans of time over which they are formed—constructed, accumulated, deposited, damaged, rebuilt, eroded, and decayed. Visual simulations can attempt to deal with this in part by being able to portray different slices of time represented at a site. This can be accomplished through a time-lapse format, or as a set or series of distinct simulations, but the full range of time, and the specific conditions at any one point in time, simply cannot be portrayed. Furthermore, limitations of time and money will often preclude the construction of a range of models. Normally, one engaged in simulating a site in virtual space will need to focus on a specific point in time, which I refer to as a *simulation slice*.[38] This is comparable to the analogy used above of a model as a slice through a ball of complex composition.

Defining the simulation slice is an important undertaking for constructing a virtual model. One must select the approximate year (or a condensed set of years), a time of the year (season), time of day (affects lighting and shadows), weather conditions, and human involvement (no humans, cutout humans, animated or not, activities appropriate to the time and place, and so on). The result is a simulation slice that cannot and should not be regarded as anything more that what it is—a representation that reflects reasoned interpretation of how a particular place may have looked, and the kinds of things that may have happened there, at some slice in time. Even so, the problem, of course, is that the methods of archaeology do not allow us the precision to chronologically differentiate all structures, features, materials, and other traces at a given site.

As an example of a simulation slice, in 2004 colleagues and I created a virtual simulation of a late prehistoric village site occupied by Mandan Indians in central North Dakota as an edutainment video for a general-public audience.[39] The site, called On-a-Slant, had been occupied for roughly 200 years, so it had undergone considerable change over the time of its occupation, as well as subsequent to its abandonment. We were contracted to create a single, three-dimensional (3D), stereo visualization that would capture the sense of place prior to the time of its abrupt abandonment. We selected as our simulation slice a time when the village was at the height of its prosperity: the year AD 1776, afternoon on a sunny day in late summer. This specificity was given in order to provide the modelers with a vision and the audience with a clear sense of time

[36] E.g., Karen M. Kensek et al., "Fantastic reconstructions or reconstructions of the fantastic? Tracking and presenting ambiguity, alternatives, and documentation in virtual worlds," *Automation in Construction* 13 (2004); Pollini et al. "Problematics of Making Ambiguity Explicit in Virtual Reconstructions: A Case Study of the Mausoleum of Augustus." In, *CHArt Conference Proceedings, volume eight, 2005. Theory and Practice* (A. Bentkowska, T. Cashen, and H. Gardiner. The British Academy, 10-11 Nov. 2005. Computers and the History of Art, 2007, http://www.chart.ac.uk/chart2005/papers/pollini.html); Sorin Hermon and Joanna Nikodem, "3D modeling as a scientific research tool in archaeology." In, *Layers of Perception. Proceedings of the 35th International Conference on Computer Applications and Quantitative Methods in Archaeology (CAA), Berlin, Germany, April 2-6, 2007* (A. Posluschny, K. Lambers, and I. Herzog, eds., paper on CD ROM. Bonn: Dr. Rudolf Habelt GmbH, 2008).

[37] Earl, 2006, 199.

[38] Clark, Jeffrey T. et al., "On-a-Slant: A Mandan Village in Virtual Time and Space." In, *Interdisciplinarity or The Best of Both Worlds, VAST2004* (K. Cain, Y. Chrysanthou, F. Niccolucci, D. Pletinckx, N. Silberman, eds., pp. 57-8. Brussels: EPOCH, 2004).

[39] Clark et al., 2004; Bergstrom et al. 2007; Aaron Bergstrom et al., "On-A-Slant Virtual Village: Constructing a 3D Stereo Imaging Exhibit." In, *The World Is in your Eyes. Proceedings, Computer Applications and Quantitative Methods in Archaeology, Tomar, Portugal, March 21-24, 2005* (A. Figueiredo and G. L. Velho, eds., pp. 61-4. Tomar, Portugal: Associação para o Desenvolvimento das Aplicações Informáticas e Novas Tecnologias em Arqueología, 2007). For a non-stereo (one eye) verison of the rendered video, go to http://onaslant.ndsu.edu/videoclips/index.htm.

and space even though such specificity in an evidential sense is completely impossible. In the absence of reliable data on all houses, we took all earth-lodge houses for which there was evidence (excluding examples of overlapping houses suggesting rebuilding), and simulated them in the village, suggesting contemporaneity. While contemporaneity of all houses is unlikely, it is no less accurate than arbitrarily selecting some number of houses to exclude. The guiding principle was to provide visual model that was appropriate to the purpose for which it was intended.

In any given model, there will be ambiguity and uncertainty. We will never have all of the evidence needed to make a model with complete precision and accuracy. Virtual simulations are always based on interpretations of multiple types and from multiple sources of information. By its very nature, interpretation is ambiguous and uncertain. Further, as there are alternate interpretations of most pieces of information, the alternate interpretations of the whole geometrically expand as the numerous pieces of evidence are brought together. This potential weakness actually leads us to the real strength of virtual modeling: through virtual models, one can much more readily present alternate "visions" or stories of the past that can serve as valuable tools for both research and education.

7 Conclusion: So, what's in a name?

The notion that archaeologists can engage in reconstruction of the past is one that developed early in the history of North American anthropological archaeology, although it is not unique to American archaeology. Despite changing perspectives on the nature of the discipline, the conception of reconstruction as central to archaeology has remained an unfortunate legacy. As Walter Taylor observed over half a century ago, to insist on attempting to reconstruct the past is to be doomed to failure. New sites, new evidence, new techniques for extracting evidence, and new perspectives on what archaeologists should be doing all inevitably mean that our "knowledge" of the past is always provisional. Moreover, there is no single model or interpretation (let alone reconstruction) of the past, but multiple possible models, depending on the perspectives taken, the complexity of the model, and the level of detail deemed appropriate for its purpose.

Instead of reconstructing the past, archaeologists are always dealing with models, or constructions. This is most evident today when we look at the domain of virtual archaeology. A virtual simulation provides a visual model of the way some place, at some simulation slice in time, may have looked. When that model is made highly realistic in appearance, it probably conveys to the viewer, more than a textual description, a sense that the model is indeed an accurate "reconstruction." The fact is, some parts of the model are more accurate than others; at one end of the range, some elements are likely to be dead on and based on abundant data, while at the other end of the extreme, some elements are purely conjectural.

Nevertheless, what many have claimed to be problems with virtual archaeology are, in fact, not problems due to the visual detail of simulations, or with the general practice, but with the implication of what they represent. As long as they are called "reconstructions" they will be viewed as in some way misleading and therefore undesirable or to be used with caution. The problem, as conceived by some, is that they can illustrate more than is known with certainty. The fact is, however, that any representation called an archaeological *"reconstruction"* whether verbal, as in conventional archaeology, or simulation, as in virtual archaeology, suffers from the same inherent problems: it purports to be what it cannot be. With visual virtual simulations, we are simply more aware of the shortcomings in data that drive it because of the apparent completeness of the representation. The virtual simulation, as a compelling visualization, is, therefore, damned by its own success.

So, what's in a name? Potentially, there is quite a lot. On the one hand, something cannot be changed from what it is by calling it something that it is not—a rose by any other name. On the other hand, by employing inappropriate terms we can be misleading, to ourselves as well as others, as to what the precise enterprise is and should be.

Bibliography

Barceló, Juan A. "Visualizing what might be: An introduction to virtual reality techniques in archaeology." In, *Virtual Reality in Archaeology*. J. A. Barceló, M. Forte, and D. H. Sanders, eds., pp. 9-35. Oxford: BAR International Series (843), 2000.

Bergstrom, Aaron, Clark, Jeffrey T., Snider, Douglas G., Frovarp, Richard, Reetz, Dan, Slator, Brian M., Landrum, James E. III, and White, Ryan. "On-A-Slant Virtual Village: Constructing a 3D Stereo Imaging Exhibit." In, *The World Is in your Eyes. Proceedings, Computer Applications and Quantitative Methods in Archaeology, Tomar, Portugal, March 21-24, 2005*. A. Figueiredo and G. L. Velho, eds., pp. 61-4. Tomar, Portugal: Associação para o Desenvolvimento das Aplicações Informáticas e Novas Tecnologias em Arqueología, 2007.

Binford, Lewis R. "Archaeology as anthropology." *American Antiquity* 28 (1962): 217-25.

Binford, Lewis R. "Archaeological Perspectives." In, *New Perspectives in Archeology*. S. Binford and L. R. Binford, eds., pp. 5-32. Chicago: Aldine Publishing CO, 1968.

Binford, Lewis R. and Sabloff, Jeremy A. "Paradigms, systematics and archaeology." *Journal of Anthropological Research* 38 (1982): 137-53.

Chamberlain, Thomas C. "The method of multiple working hypotheses." *Science* 15 (1890): 92-6.

Clark, J. Grahame D. *Star Carr*. Cambridge: Cambridge University Press, 1954.

Clark, Jeffrey T. and Terrell, John. "Archaeology in Oceania." *Annual Review of Anthropology* 7 (1978): 293-319.

Clark, Jeffrey T., Snider, Douglas G., Bergstrom, Aaron,

Frovarp, Richard, Reetz, Dan, Slator, Brian M., and White, Ryan. "On-a-Slant: A Mandan Village in Virtual Time and Space." In, *Interdisciplinarity or The Best of Both Worlds, VAST2004,* K. Cain, Y. Chrysanthou, F. Niccolucci, D. Pletinckx, N. Silberman, eds., pp. 57-8. Brussels: EPOCH, 2004.

Clarke, David (ed.). *Models in Archaeology*. London: Methuen, 1972.

Earl, Graeme P. "At the edges of the lens: Photography, graphical constructions and cinematography." In, *Digital Archaeology. Bridging Method and Theory,* T. L. Evans and P. Daly, eds., pp. 191-209. London: Routledge, 2006.

Eiteljorg, Harrison II. "The Compelling Computer Image - a double-edged sword." *Internet Archaeology,* Issue 8 (2000). http://intarch.ac.uk/journal/issue8/eiteljorg_toc.html

Flannery, Kent V. "The golden Marshalltown: a parable for the archaeology of the 1980s." *American Anthropologist* 84 (1982): 265-78.

Forte, Maurizio. "About virtual archaeology: Disorders, cognitive interactions and virtuality." In, *Virtual Reality in Archaeology*. J. A. Barceló, M. Forte, and D. H. Sanders, eds., pp. 247-59. BAR International Series (843), 2000, Oxford: BAR Publishing.

Forte, Maurizio and Siliotti, Alberto (eds.). *Virtual archaeology: Re-creating Ancient Worlds*. New York: Harry N. Abrams, Inc., Publishers, 1996.

Gibbon, Guy. *Anthropolgical Archaeology*. New York: Columbia University Press, 1984.

Hempel, Carl. *Philosophy of Natural Science*. Englewood Cliffs, NJ: Prentice-Hall, Inc., 1966.

Hermon, Sorin and Nikodem, Joanna. "3D modeling as a scientific research tool in archaeology." In, *Layers of Perception. Proceedings of the 35th International Conference on Computer Applications and Quantitative Methods in Archaeology (CAA), Berlin, Germany, April 2-6, 2007*. A. Posluschny, K. Lambers, and I. Herzog, eds., paper on CD ROM. Bonn: Dr. Rudolf Habelt GmbH, 2008.

Hodder, Ian. "Theoretical archaeology: a reactionary view." In, *Symbolic and Structural Archaeology,* I. Hodder, ed., pp. 1-16. Cambridge: Cambridge University Press, 1982.

Hodder, Ian. *Reading the Past: Current Approaches to Interpretation in Archaeology*. Cambridge: Cambridge University Press, 1986.

Hodder, Ian (ed.). *The Archaeology of Contextual Meaning*. Cambridge: Cambridge University Press, 1987.

Hodder, Ian. "Interpretive Archaeology and its role." *American Antiquity* 56 (1991): 7-18.

Hodder, Ian, Shanks, M., Alexanrir, A. Buchli, V., Carman, J. Last, J., and Lucas, G. (eds.). *Interpreting Archaeology. Finding Meaning in the Past.* London: Routledge, 1995.

Kensek, Karen M., Lynn Swartz Dodd, and Nicholas Cipolla. "Fantastic reconstructions or reconstructions of the fantastic? Tracking and presenting ambiguity, alternatives, and documentation in virtual worlds." *Automation in Construction* 13 (2004): 175-86.

Kroeber, Alfred. L. "Archaeology." In, *Encyclopedia of the Social Sciences, Vol. 2*. New York: The Macmillan Co., 1937.

Miller, P. and Richards, Julian. "The good, the bad, and the downright misleading: archaeological adoption of computer visualization." In, *Computer Applications and Quantitative Methods in Archaeology 1994*. J. Huggett and N. Ryan, eds., pp.19-22. Oxford: BAR International Series (600), 1995.

Pollini, John, Dodd, Lynn Swartz, Kensek, Karen, and Cipolla, Nicholas. "Problematics of Making Ambiguity Explicit in Virtual Reconstructions: A Case Study of the Mausoleum of Augustus." In, *CHArt Conference Proceedings, volume eight, 2005. Theory and Practice*. A. Bentkowska, T. Cashen, and H. Gardiner, eds. The British Academy, 10-11 Nov. 2005. Computers and the History of Art, 2007. http://www.chart.ac.uk/chart2005/papers/pollini.html

Reilly, Paul. "Towards a virtual archaeology." In, *Computer Applications in Archaeology 1990*. K. Lockyear and S. Rahtz, eds., pp. 133-39. Oxford: British Archaeological Reports (Int. Series 565), 1990.

Renfrew, Colin. "Towards a cognitive archaeology." In, *The Ancient Mind. Elements of Cognitive Archaeology*. C. Renfrew and E. B. W. Zubrow, eds., pp. 3-12. Cambridge: Cambridge University Press, 1994.

Ryan, Nick S. "Computer based visualisation of the past: technical 'realism' and historical credibility." In, *Imaging the Past: Electronic Imaging and Computer Graphics in Museums and Archaeology*, T. Higgins, P. Main, and J. Lang, eds., pp. 95-108. London: British Museum Press (British Museum occasional paper 144), 1996.

Sanders, Donald. H. "Archaeological publications using virtual reality: case studies and caveats." In, *Virtual Reality in Archaeology*. J. A. Barceló, M. Forte, and D. H. Sanders, eds., pp. 37-52. BAR International Series (843), Oxford: BAR Publishing, 2000.

Sanders, Donald. H. "Why Do Virtual heritage?" In, *Digital Discovery: Exploring New Frontiers in Human Heritage. CAA 2006. Computer Applications and Quantitative Methods in Archaeology. Proceedings of the 34th Conference, Fargo, United States, April 2006*. J. T. Clark and E. M. Hagemeister, eds., pp. 427-36. Budapest: Archaeolingua, 2007.

Schiffer, Michael B. *Behavioral Archeology*. New York: Academic Press, 1976.

Schlereth, Thomas J. "Material Culture Studies in America, 1876-1976." In, *Material Culture Studies in America: An Anthology*. T. J. Schlereth, ed., pp. 1-78. Walnut Creek, CA: AltaMira Press, 1982.

Shanks, Michael. "Listening to things: augmented reality and the science question in archaeology. The Science Question in Archaeology" (2006). Last modified: Dec 20, 2006. http://documents.stanford.edu/michaelshanks/77

Shanks, Michael and Tilley, Christopher. *Re-constructing Archaeology*. Cambridge: Cambridge University Press, 1987.

Shimkin, Dimitri. "Models for the Downfall: some ecological and culture-historical considerations." In, *The Classic Maya Collapse*, T. P. Culbert, ed., pp 269-99. Albuquerque: University of New Mexico Press, 1973.

Steward, Julian H. "The direct historical approach to archaeology." *American Antiquity* 7 (1942): 337-43.

Sylaiou, Stella and Patias, Petros. "Virtual reconstructions in archaeology and some issues for consideration." Imeros vol 4 (2004). Accessed April 21, 2009. http://www.ime.gr/publications/print/imeros/en/04/index.html

Taylor, Walter W. *A Study of Archeology*. Memoir No. 69. *American Anthropologist* 50(3), Part 2, 1948.

Taylor, Walter W. "Old wine and new skins." In, *Contemporary Archaeology: A Guide to Theory and Contributions*, M. P. Leone, ed., pp. 28-33. Carbondale: Southern Illinois University Press, 1972.

Thomas, Julian. *Interpretive Archaeology. A Reader*. Continuum International Publishing Group, 2001.

Tilley, Christopher (ed.). *Interpretative Archaeology*. Oxford: Berg, 1993.

Watson, Patty Jo, Le Blanc, Steven A., and Redman, Charles. *Archaeological Explanation: The Scientific Method in Archaeology*. New York: Columbia University Press, 1984.

Wedel, Waldo R. *The Direct-Historical Approach in Pawnee Archeology*. Washington, D.C.: The Smithsonian Institution, 1938.

Willey, Gordon R. *Prehistoric Settlement Patterns in the Virú Valley, Peru*. Washington D.C.: Smithsonian Institution, Bureau of American Ethnology, 1953.

Willey, Gordon R. and Phillips, Philip. *Method and Theory in American Archaeology*. Chicago: University of Chicago Press, 1958.

Exploring cognitive landscapes: toward an understanding of the relationship between space/time conceptualization and cultural material expression

J. van der Elst[1]
[1]University of New Mexico, Albuquerque, NM, USA.

Abstract

Indigenous histories are known through research within the related disciplines of anthropology and archaeology, but disagreement often exists regarding the interpretation of that history with Indigenous scholars and communities, rooted in epistemological differences. Geospatial technologies, which are increasingly employed to explore human spatial and cultural material relationships, have developed based on the idea that human spatial cognition is universal, therefore the underlying framework for human spatial experience is the same or similar. However, it has been shown in recent studies in linguistics that significant differences exist between language groups in this fundamental cognitive domain, warranting research into how space/time is perceived, understood, and represented across the human population. A case study of the ancestral Pueblo culture of the Northern Rio Grande region, New Mexico is used to investigate the usability of geospatial technologies for cross-cultural understanding and addresses related methodological issues. Additionally, the paper synthesizes research results from different perspectives in an effort to expand our ability for understanding human experience.

***Key words**: spatial cognition, epistemological diversity, cross-cultural understanding*

1 Introduction

The history of Native American culture has been studied within the disciplines of anthropology and archaeology. Archaeology, the discipline of things, studies material cultural remains of past humans. Things as cultural objects are of interest, but it is the relationship between things and the location of these in space that tells us how these things were used and the role they played within a larger organization of space, geographical, social, or other. That is, archaeologists infer past socio-economic organization based on patterns of cultural remains observed in the present landscape, interpreted within a certain conceptualization of space and time. While the inferential process differs between schools of thought, most archaeological data stem from empirical observation of things and patterns of things, and the retrieval and observation of material remains through field methods of survey and/or excavation. What are material things is seldom controversial; they belong to the physical domain and can be observed and perceived through our senses, most importantly visual and tactile. Knowledge is then obtained through the study of material remains by standard scientific practices.

Indigenous scholars have articulated metaphysical and epistemological differences that exist between Indigenous and Western philosophy. As the foundation for modern science practice, Western philosophy or philosophy of science underpins all scientific knowledge construction. These differences in general refer to the fact that Indigenous knowledge is the result of synthesis and focus on processes and relationships, more so than analysis. Place and geographical connection are paramount, however the underlying conceptualization is not necessarily based on a Cartesian framework (Burkhart 2004, Cajete 2004, Smith 1999, Waters 2004, Whitt 2004).

To investigate the idea or the possibility of differences in spatial cognition two lines of inquiry are pursued in addition to the Indigenous philosophical discourse, first the developments in GIScience over the last two decades and second, recent research in psycholinguistics (see figure 1). Geospatial technologies, such as Geographic Information Systems (GIS) and Remote Sensing (RS) are increasingly used within the social sciences; critique of these technologies as being overly environmentally deterministic has led to substantial changes in research direction (Goodchild 1992). A new direction in GIScience started out as Critical GIS (Kwan and Lee 2004, Schuurman 2002), based on the idea that space is not experienced by all people in the same way, such that general spatial (social) categories are not suitable in all situations. Most research along these lines has focused on spatial ontologies; how humans experience space, qualitative spatial reasoning, is gaining interest over understanding space quantitatively.

Related to this research are current findings in the field of psycholinguistics, that have shown that significant differences in human spatial cognition exists in coordinate systems and spatial categories used between speakers of different language groups and the connection between language and human thought (Levinson 2003, Levinson and Wilkins eds. 2006). This holds important implications for the spatial sciences, as well as for archaeological research through which interpretations of cultures are derived from observed patterns of material cultural remains.

The case study is specifically focused on ontology, the definition and classification of entities and the relationships between these that underpin the principles of reality. Ontological and epistemological questions concern what is commonly referred to as a person's worldview. For instance, in an objectivistic worldview (or paradigm, Creswell.and Plano-Clark 2007), a mountain is a mountain for everyone, that is, the meaning of a phenomenon is inherent to the phenomenon and can be experienced by interacting with it (Lakoff and Johnson 1980). However epistemological and related ontological differences exist between Western and Non-western ways of thinking as well as within the Western scientific community. In order to provide a context for the case study two issues need to be addressed, first, what is the nature of the epistemological differences described by Indigenous scholars, and second, can different ways of world conceptualization, or worldviews, be explored and communicated employing currently available geospatial technologies.

2 BACKGROUND

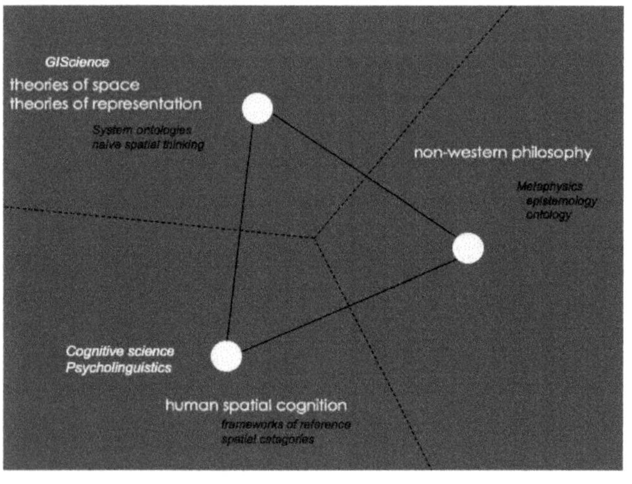

Figure 1. Outline of Theoretical Framework.

Inherent in the current organization of the sciences is a reductionist approach to many research problems. The classification of the sciences and the philosophy of science as conceived by Comte holds that the sciences build knowledge in a hierarchical, nested manner; sociology (and anthropology) are placed at the top and integrate knowledge about the world and the role of humans within it, resulting in a positivist research paradigm. The positivist and subsequent neo-positivist and post-positivist paradigms have been criticized based on the underlying assumption in Comte's original conception, namely that the (material) world is known through the natural sciences and through the study of its parts we can arrive at an understanding of the whole (Giddens 1974, Habermas 1984, Kuhn 1970 Wallerstein 2004). Kuhn's suggestion that the evolution of science was at least in part determined by social factors, opened up new research directions questioning objectivity and validity of science practice, resulting in two opposing, (but not new) viewpoints of reality with associated epistemologies. In its extreme forms these can be summed up as: one, an independent reality exists, it can be known through logic, reasoning and empirical observation, which is uncontested as we can all observe that same reality, it is absolute. Two, on the other side of the spectrum reality does not exist outside our construction (in our minds) of it; it is relative and dependent upon the observer.

A change in the current research landscape is occurring with researchers who no longer adhere to one viewpoint or the other and seek to go beyond disciplinary boundaries; such researchers are small in numbers but growing. Wallerstein refers to these research efforts as either complexity or cultural studies. Complexity studies, which are inherently interdisciplinary, are based on the idea that the whole is greater than the sum of its parts and seek to understand emergent properties that are the result of non-linear dynamic forces, which are difficult to solve analytically. Consequently requiring a different methodology, one in which a top-down approach goes hand in hand with a bottom-up approach, and synthesis as well as analysis are necessary components for understanding. (DeLanda 2000). One such research network is centered on the importance of space and place, as many researchers have realized the importance of spatial aspects in their research (Manson and O'Sullivan 2006), as well as the argument that space and time are fundamental concepts in human thinking and reasoning (Lakoff and Johnson 1980). Space (and place), originally the research domain of the discipline of geography has ignited interesting discussion and developments, notable research under the umbrella of Geographic Information Science (GIScience) (Goodchild 1992).

The broader philosophical context for this research addresses the fundamental issue of ontology and is concerned with the questions of what are space and time and how do we know about space and time and spatial reasoning, The questions have long been central to philosophical discourse within Western thinking/science and their discussion is beyond the scope of this paper, but are extensively discussed elsewhere (see Puequet 2002, Raper 2000). These ideas underlie the developments of geospatial technologies, including those that are focused on understanding how humans experience spaces, although certain theories of spatial reasoning are rooted in different underlying concepts of nature (Whitehead 1919).

In efforts to understand spatial reasoning it is mostly assumed that human spatial cognition is universal affirmed by spatialization evident in language through the use of spatial metaphors (Lakoff and Johnson 1980). This universality assumption is also perpetuated by the fact that many empirical studies, based on this assumption, employ subjects that are drawn from a select pool, mostly western, urban societies (Levinson 2003, Nesbitt 2003). Non-western philosophies however, present diverging viewpoints as they suggest that space and time as conceptualized in Western philosophy may not be universally fundamental in human thinking (Smith 1999, Hester Jr. 2004).

GIScience: Systems of representation, such as GIS, can be employed to analyze material things and relationships and synthesize these within a larger spatial context. In order to do this, the existence of a physical reality, within which these things are placed and relate to other objects, and the fact that we can know and thus represent this reality, needs to be accepted, according to Raper (Fisher and Unwin eds. 2005, Raper 2000). Theories of space, formalizing the nature of matter and void, underpin these representational systems. Representational systems however are abstractions at best and choices are made in the design of these systems as to how and what is represented to provide a (conceptual) model that can help us understand the world around us. Critical GIS questions western representation and categories of space, hypothesizing that such frameworks and categories support a dominant view and may not be true for all of human –spatial- experience. This critique is especially important for socio-cultural research (Kwan and Lee 2004), and has led to a broader research direction into ontologies of space (Schuurman 2002, Mark, Turk, and Stea 2007). Notably, how spatial information systems can move beyond geometric representation to ones that are qualitative, in order to understand human spatial experience, movement, and navigation (Freska et.al. eds. 2003, Galton 2000). To do so requires an investigation of how objects and their (spatial) relationships are conceptualized within a larger cosmology.

Spatial concepts and theories have been derived from mereotopology, which combines the disciplines of mereology and topology. Mereology is the theory of parts and wholes. The entities whose parts form the subject matter of mereology are variable, including abstract and physical entities (Galton, 2000: p 70). As a theory of spatial reasoning, however, mereology lacks a key component, i.e. the notion of connection or relationship, therefore topological notions are added to provide a qualitative description of space (Galton, Whitehead). Within the geospatial sciences the suitability of mereotopology is for example explored by Mark and Frank (1991), in an approach defined as ethnophysiography, and Bibby (2005). In addition, analysis of natural language has provided insight in object-categorization, for instance topological notions and conceptual structuring can be illustrated through the use of prepositions. In representations of geographic space an additional problem is related to the nature of geographic entities itself that are currently used, many of which have no clear boundaries. (Casati and Varzi 1999, Galton 2000, Smith 1996). Natural language however is not the only source for investigation categorization and conceptualization. A general ontology, as argued by Smith, would draw from "other cognitive modes of access to reality, including perception, scientific theories, the map-making activities of the geographer, knowledge-sharing systems, and so on (Smith 1996: p295)." It is within the discipline of linguistics however that most of the information regarding the diversity of human spatial cognition is generated (Levinson 2003, Smith 1996).

Linguistics: As argued by Levinson, the Kantian idea of a universal framework for human thinking, an *a priori* spatio-temporal framework to which we 'map' our experiences that assumes man as the center and is egocentric, is a framework that underlies many scientific efforts and representations of space and time. It provides the foundation for phenomenology and support for the ideas of space as fundamental, which also comes from the spatialization in our language (Levinson 2003,Lakoff and Johnson 1980). If different underlying frameworks are possible, as suggested by non-western philosophical discourse, it does not necessarily negate Kant's ideas about knowing and knowledge construction, it merely indicates that the way the world can be known based on human sense-experience and reasoning is much broader than previously thought.

The problem with current systems of interpretation and representation is that they are based on a narrow idea of underlying framework that mimics human thinking. Indication that significant differences in human spatial cognition exist comes from an empirical research program of the Max Planck Institute (MPI) in psycholinguistics. This program, led by Levinson, is characterized by many separate research efforts sharing the same methodology, enabling a cross-linguistic comparison at a scale rarely ever accomplished. It is loosely based on the Sapir-Whorf hypothesis of linguistic relativity that lately has gained renewed interest. This research program has shown that significant differences exist between different language groups across spatial subdomains, indicating that differences exist in the use of spatial coordinate frameworks (directional) (i.e not all groups use a relative frame of reference!) but also that there seem to be no universal landscape categories, such as mountain and river, as indicated by toponymic research. These differences cannot be attributed to environment or subsistence strategy alone (Majid, Enfield, van Staden eds. 2006). The consequence is that cognitive research in the spatial domain, thus far almost exclusively based on western test subjects (Indo-European language speakers), possibly only represents a fraction of the human spatial cognitive potential. In a similar vein, the focus on the visual has been argued to represent a Western preference. This research program has developed several novel ways of investigating differences in human spatial cognition. Levinson's method of category research identifies three main factors that influence the construction of landscape categories and placenames: perceptual salience, that would be indicated by the existence of universal categories, affordances (sensu Gibson 1977), and cultural ideas. Within his study he found that landscape categories are most influenced by affordance and cultural ideas. (Levinson 2008, see for another approach: Regier 1996)

Native philosophy: Smith has discussed the difference in conceptualization between Western and Indigenous thinking and research methodologies, especially as it has been affected by colonization. The notion of moral behavior not as a societal construct but as an innate (universal) human quality suggested by Rousseau has developed within western philosophy of science and later theories of social evolution as unilineal development of human thinking (Smith 1999). The first of such theory formalized by Comte in the three stages (theological, metaphysical, and positivist) underpins modern science. Colonialism however confronted European thinkers with different ideas and worldviews, but conveniently, the idea of unilineal evolution could be exploited to justify economic and political domination, allowing Europeans to argue that the indigenous people they encountered were less developed. In this way scientific inquiry and political/ideological interest has stayed intertwined, represented and reified for instance through cartographic practice and commodification of knowledge (Habermas 1984, Root 1993, Whitt 2004). Because these different knowledge systems were hypothesized to be simple, early stages of human development, they could be studied to understand classification systems and modes of thought (Smith 1999: p 50). With regard to the concept of space and time, Smith notes that western ideas about space and time are encoded in language, philosophy and science, but that this does not hold for a number of indigenous languages.

Other fundamental concepts or frameworks underlie many non-western worldviews. For example, Burkhart describes a Native American epistemology as lived experience; this pertains to a more general knowledge that can be called synthesis, incorporation or understanding. A native philosophical understanding, akin to a phenomenological approach, must include all experience, not simply one's own, all experience that has passed down in stories (Burkhart 2004: 16-25). However, a phenomenological approach is suitable when the underlying worldview on which all these experiences are 'mapped' is the same or similar; understanding human experience and expression that is based on a different conceptual framework or metaphysics is much more complicated (Birnbaum 2008). An important underlying principle in Native American philosophy is relatedness, for instance: all things in the universe are related to air and are all made up of the same basic element. Flow, movement, or change, as well as interstitial spaces are mentioned as fundamental underlying concepts, (compared to Space and Time as fundamental in western philosphy) indicating a complex worldview (Cajete 1999,2004, Cordova 2004, Jojola 2004, Smith 1999, Swentzell 1990, Waters 2004).

In summary it can be said that scholars can no longer assume that spatial cognition is universal, therefore new ways of understanding spatial reasoning need to be developed. Based on an understanding of the epistemological differences that exist between western and non-western ways of thinking, it needs to be investigated if these differences can be addressed within current scientific research frameworks, focused on space and time. Epistemological foundation of Native American philosophy is based on synthesis and concepts of relatedness, process and flow. This is different from Western scientific traditions that are based on concept of space and time and reductionist approaches. Only recently have other ideas begun to be explored, challenging the epistemological and ontological foundation of scientific research (DeLanda 2000, Manson and O'Sullivan 2006). As argued by Manson and O'Sullivan the epistemological and ontological foundation of systems based on the complexity is flexible and holistic, however its dominant epistemology is based on computational modeling driven by the need to "focus on entities and their relationships" across multiple scales. Challenges in employing complexity models in space and place-based research, such as verification and validation, the linkage between pattern and process, among other things are discussed elsewhere (DeLanda 2000, Manson and O'Sullivan 2006) and are currently beyond the scope of this paper.

In short, the research directions in the spatial sciences overlap in efforts toward understanding epistemological and ontological foundations for human experience. This intersection provides a basis for exploratory research investigating if current geospatial technologies can provide a basis for different interpretive frameworks in archaeology.

3 CASE STUDY: EMPLOYING GEOSPTIAL TECHNOLOGIES TO UNDERSTAND CULTURAL EXPERIENCE

The case study is focused on the ancestral Pueblo culture of the Northern Rio Grande region, New Mexico and investigates if current interpretive frameworks used in archaeological research are suitable. The research is based on the above-discussed differences between Western and Indigenous philosophies. Archaeologists have defined the history of human nature in epochs of technological inventions, each new tool used to control nature. In that tradition, archaeologists use these new digital tools in the Information Age to find and control the past (Zubrow 2006). Even though, work that is currently being done as digital archaeology is more nuanced than this simple statement may imply, as is thoughtfully discussed by Zubrow, the fact is that representations of the past are reified through the use of digital and virtual technologies. It is therefore crucial to question the methodologies and underlying conceptual frameworks that archaeologists use because they may not necessarily represent the worldview of the culture that is studied.

Geospatial technologies include Geographic Information Systems (GIS) and Remote Sensing (RS) technologies. Within current representational and analytical systems two main data models are in use, vector and raster data models. The vector model is used primarily to represent discrete features and can have a point, line, or polygon geometry. The raster model is better suited to represent continuous surfaces, such elevation surfaces, geographic phenomena that do not have clear spatio-temporal boundaries, for example mountains. To explore the concepts and categories used in Puebloan metaphysics the raster model is preferred. Others have experimented with the use of raster-based modeling systems and agent-based models in archaeology, which allows for multi-dimensional modeling integrating environmental data and human decision making algorithms (Kohler and van der Leeuw eds. 2007, Gimblett ed. 2002). This research however, takes a different approach, as it is assumed that decision-making rules are at least partly based on worldview. Exploring this worldview through investigating underlying concepts and categories is suggested to provide a basis for cross-cultural understanding and system development that can support different epistemologies for research and educational purposes.

Much of the current information of the earth's surface is gathered through remote sensing systems, that generate aerial and satellite imagery that represents values collected for certain parts of the Electro Magnetic spectrum. In short, sensors can for instance collect data in the visible range, measuring reflectance values of entities and phenomena as determined by their physical and chemical properties. For example, changes in vegetation health can be monitored by assessing changes in 'greenness' over time. Greenness, or more correctly, healthy greenness, in this context can be considered an object. The science of remote sensing also focuses awareness of the importance of the atmospheric and hydropsheric processes in the biosphere. Through image processing techniques of multidimensional data it is also possible to visualize aspects, entities, and processes that are not (easily) detected by the human eye and has greatly enhanced our understanding of ecological processes that occur at different spatial scales and temporal cycles. The fact that certain aspects cannot be sensed through our visual system does however not mean that we cannot sense it in other ways, through smell, hearing or touch for instance. The visualization (or made sensible through other methods such as sonification), of these data however, makes it possible to understand these processes and the connection between things in new ways, as well as look at objects in a different way as understood through the relationships and processes. Remote sensing principles and the raster data model provide a stepping-stone for exploring concepts and categories of space. *The Ethnogeography of the Tewa Indians*, which includes placenames as well as categories recorded by Harrington, is used as a dataset (Harrington 1916).

Today Pueblo communities in New Mexico, descendant communities of people who have lived in this region for centuries, occupy 19 pueblos. Pueblo communities are autonomous and the different languages spoken are highly diverse, containing three distinctive language groups, Zunian, Keresan, and Tanoan, that exhibit dialectical differences (Dozier 1983).

Speakers of the Tewa language today live in six of the 19 New Mexican Pueblos (Mithun 1999,Ortiz 1969). Ancestral cultural material of these communities is omnipresent, visible on the modern day landscape and is part of the ongoing Tewa world experience. The conceptual world of the Tewa specifically has been discussed in several publications, notably by Ortiz and Dozier and general characteristics of American Indian metaphysics underlie this worldview (Cajete 1999 Dozier 1983 , Ortiz 1969, Waters 2004) Dozier states that in general the origin myths of all Pueblos are similar. In addition to a shared concept of origin, concepts regarding the nature of the universe (metaphysics), conceptual ordering and spatial arrangement of the pueblos is also shared. In this framework, the Pueblo is the center of the universe. Several Indigenous researchers have written about the importance and the place of the Pueblo in the Puebloan cosmology. Ortiz for instance describes the Tewa world as existing of a number of zones with the Pueblo at its center; the surrounding physical landscape is tightly integrated with subsistence activities and the ritual calendar, thus maintaining a relationship with the larger world, horizontal as well as vertical. This interconnectedness between the pueblo and the surrounding world, is for example reflected in the pueblo village layout, which exist of four parts and maintains open corners (Ortiz 1969:128)

Several authors have discussed, notably Swentzell, that there is a fundamental difference between puebloan design and western design principles. The goal of and of human existence within the puebloan design is to maintain wholeness with the natural universe (instead of conquer and control), people are not separate from nature; this is reflected clearly in puebloan architectural design and evident from the subtle or inconspicuous markers that reflect Tewa cosmological organization, as outlined clearly by Ortiz. (Cajete 1999, Ortiz 1969, Swentzell 1997). As described by Swentzell, people live at the center of their universe and the purpose of life is to be united with nature, connected with everything in the natural world. Directional forces, rather than being upward as in Christian ideology, are cyclical and move in and out of the earth. This flow of life or *Po-wa-ha*, (water-wind-breadth) symbolized by the movement of water and wind determines existence rather than a physical body or other physical manifestation.

Often scientists operate under the assumption that the objects of study and taxonomies used to order these objects, correspond to natural kinds in nature, that is, the grouping or ordering (of objects, of properties, or characteristics) does not depend on humans. This real and independent existence justifies scientific inference and practice (SEP, accessed October 14,2009). Object in philosophy refers to a thing, an entity or a being, but this can have multiple meanings. In the broadest sense, an object can be anything that we think or talk about , or it can refer to things that have properties and relationships to other things, but are contrasted with them, in the most restricted sense objects only refer to physical bodies located in space and time, material bodies, or simply inanimate matter. What are things and how things are classified is dependent on the metaphysical framework used and its associated ontology.

For example Galton (2000) discusses several object classes and the principles by which these object classes can be organized. Objects that possess spatial attributes, he argues, must be dependent in some way on matter. This dependence forms one of the organizing principles to classify objects in three kinds of objects: material, parasitic, and perspectival. Material objects refer for instance to a piece of matter that exist in the same form over an extended period of time, this is not always straightforward as shown by his example of 'pebble' that changes form due to erosional processes. Parasitic objects, depend on existence of other objects. Examples are shadows and holes, the object of shadow for instance involves a four-place relation: x is a shadow cast by y on surface z in the light coming from w. The Pueblo component terms can be thought of as belonging to this category of objects. A Perspectival object depends for its existence on a particular point of view. A good example is the horizon, a category that involves a three-place relation: x is the silhouette of y as seen from point z. An example in Tewa is the 'great gap', *Wjio*, known to all Tewa, and at the Pueblo of Santa Clara the sun rises through this gap (Harrington 1916). It has also been documented for several languages that objects can have different names depending on the position of the viewer (Mark, Turk and Stea 2007). This classification of objects, used here as a starting point, allows the inclusion of relationships and explore 'connectedness', concepts that are noted as being fundamental in Native American thinking. The temporality of objects has much to do with cycles in nature, and related changes in relationships between objects.

Toponyms and Cosmography: The study of toponyms and landscape categories can provide insight in conceptualization and categories that underlie the patterns that are observed by archaeologists, based on the assumption of linguistic continuity of indigenous languages spoken in this region and the fact that fundamental concepts are bound up in language (Ortiz, Dozier,Waters). Toponymy, the scientific study of placeanames, despite its long history, has garnered very little theoretical interest, according to Levinson. Lately it has regained some interest and questions regarding density of names, whether or not they are descriptive have been addressed (Hunn 1996, Kari 1996). In addition to these questions it is also of interest how these places function in the larger worldview. For instance, Williams related Navajo placenames with stories, showing that placenames function in concurrence with oral stories as oral cartography used to navigate the land (Williams 2000).

The gazetteer of Tewa placenames recorded by Harrington is extensive in that it provides information on the names of places known in the different languages and dialects, including Spanish and English names, as well as origin of the placename, if known, and a description. In addition, all the places are located on accompanying mapsheets (Harrington 1916). This allows for the creation of 'knowledge maps': locations entered in a GIS that can generate different layers for each language or dialect as a raster surface of known and unknown areas as binary data set in a raster resolution that can be adjusted to any other available data. Summing these raster layers creates an accumulative 'knowledge map'. This initial exploration shows that simple geometric boundaries around distinct groups cannot be drawn. Similar findings, based on computational modeling of language diversity compared to geographic distribution of basketry forms and building styles, resulted from a study by Jordan using a variety of methods to test cultural transmission and social learning (Jordan 2007).

Before doing any kind of sophisticated density analysis, the type of placenames and function(s) it may have had needs to be better understood, as these names may represent several different, but not necessarily exclusive, networks through linguistic analysis and ecological connectedness. For instance, different seasonal, agricultural or ritual activities may be related, form stories, and/or provide a navigational function. Certain placenames provide information on plant and animal habitats and this can for instance be related to modern information derived from remote sensing data analysis. Assuming that climatological cycles are similar for modern times and the past century at least (Ni. et.al. 2002), modern landscape changes, cycles, and processes can provide additional information on how some of these names may be meaningful, such as changing rainfall patterns related to farming and field locations. In this way flexible parameters can be set for an exploratory environment. Levinson's model of perceptual salience, affordance and cultural ideas serves as a startingpoint (Levinson 2008).

So far, the data introduced can be explored within a traditional geospatial environment, related to the earth's surface from a traditional map perspective. To understand and communicate different ways of being in the world however, this framework maybe too limited, shown with a few examples below. Galton's classification of kinds of objects provides the initial framework (Galton 2000).

Locating the placenames on modern topographic maps, or georeferencing within a GIS has generated some other ideas. Many of the placenames are difficult to locate, as these are dependent on perspectives from being in the landscape. Landmarks and spatial navigation studies focused on the use of landmarks often interpret landmarks as being features that are recognized from multiple perspectives in the same way, for instance a specific building. The Tewa names often appear to be more perspectival in nature, the composition of natural elements that can be recognized from a specific perspective. It is therefore suggested that these names function is a way akin to Williams' discussed oral cartography, i.e. these landmarks cannot be automatically extracted from a Digital Elevation Model (DEM) for instance, but require at least a three-dimensional (3D) knowledge of the landscape (Berardini 2005, Williams 2000).

Other possible perspectival objects, as mentioned above, are placenames that are referred to as 'gaps'. One example worth mentioning in this context is a place referred to as 'water wind point', also known as 'water-air spirits' (Harrington 1916: 151). This point is described as a projecting point that forms a narrow gap through which the river passes and is always windy (see fig. 2). To understand the relationships expressed a different representational environment seems necessary. The landscape morphology is such that air moving through this 'gap', and air pressure produces distinctive sounds that can be interpreted in multiple ways. Understanding wind patterns, the perception of these patterns and the information these provide by being in the land cannot be easily explored using currently available GISystems, requiring other virtual /simulation systems that can visualize and facilitate other sense experiences as well.

Figure 2. Example of 'Landmark' that is viewed from a specific perspective.

A final example in this paper is related to the centrality of the Pueblo in Tewa cosmology. In most archaeological studies, the pueblo is of interest for its architecture and can be compared with other architectural achievements for providing information on socio-economic status. Within a Tewa cosmology however, it seems that the importance of the pueblo is based on its relationships with the surrounding environment. The building itself will return to the earth when no longer occupied (Swentzell). The terminology recorded for the San Ildefonso Pueblo by Harrington are based on the relationship with the sun (see fig. 3). These 'parasitical' names, sunside, shadowside, can be interpreted as evidence for the integrated function of the pueblo in cyclical patterns that underlie the many aspects of Pueblo life, ritual and subsistence activities. The form and location of architectural structures therefore should be understood through its complex relationships with many interlocking natural cycles. In addition to the placenames, or toponyms, that indicate places on the earth's surface, it is worth mentioning that the large number of terms in the cloud category for instance indicates that the earth's surface cannot be separated from vertical processes in the biosphere for understanding a Puebloan worldview (see fig. 4).

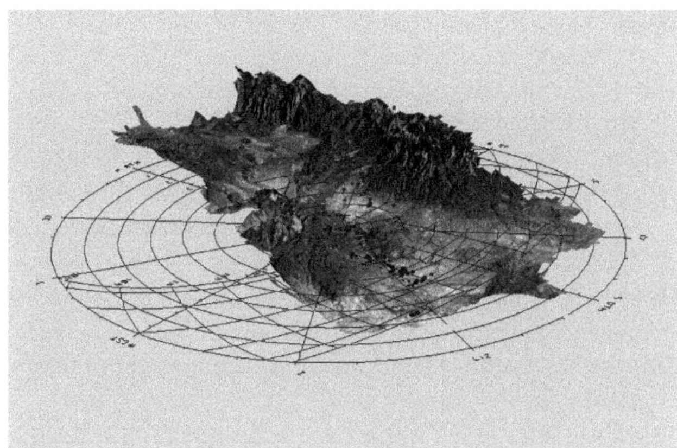

Figure 3. Exploring cycles and processes, the sunpath centerd on a specific Pueblo.

Further research:

This paper has outlined some of the issues that hinder cross-cultural understanding, in addressing these issues new ways can be explored that can inform and underlie future developments in geospatial technologies that can facilitate cross-cultural understanding and promote synergistic approaches. The intersection of research interest in diversity of human spatial cognition between the spatial sciences, linguistic research and non-western philosophies provides a starting point to move in this direction.

Figure 4. Beyond the Earth's surface, including categories for atmospheric elements.

References

Bernardini, W. *Hopi Oral Tradition and the Archaeology of Identity*. Tucson: University of Arizona Press. 2005

Bibby, P. "GIS, Worldmaking and Natural Language" in *Representing GIS,* Fisher, P., and Unwin, D. (eds). West Sussex: John Wiley & Sons Ltd. 2005:p55-83

Birnbaum, D. *The Hospitality of Presence: Problems of*

Otherness in Husserl's Phenomenology, New York: Sternberg Press. 2008

Burkhart, B.Y. What Coyote and Thales Can Teach US: An Outline of American Indian Epistemology. In: *American Indian Thought,* A.Waters, (Ed.) Malden: Blackwell Publishing. 2004: pp 15-26

Cajete, G. (ed) . *A Peoples's Ecology, Explorations in Sustainable Living,* Santa Fe: Clear Light Publishers, 1999

Cajete, G. Philosophy of Native Science. In: *American Indian Thought,* A.Waters, (Ed.) Malden: Blackwell Publishing. 2004: pp 45-56

Casati, R. and Varzi, A.C. *Part and Places, the structure of spatial representation,* Cambridge: MIT Press 1999

Cordova, V.F. Approaches to Native American Philosphy. In: *American Indian Thought,* A.Waters, (Ed.) Malden: Blackwell Publishing. 2004: pp 27-33

Creswell, J.W. and Plano Clark, V.L. *Designing and Conducting Mixed Methods Research.* London: Sage Publications 2007

De Landa, M. *A Thousand Years of Nonlinear History*. New York: Swerve Editions. 2000 [1997]

Dozier. E. P. *The Pueblo Indians of North America,* Illinois: Waveland Press. 1983

Fisher, P., and Unwin, D. (eds). *Representing GIS,* West Sussex: John Wiley & Sons Ltd. 2005

Freska, C., Brauer, W., Habel, C., and Wender, K.F. (Eds.) *Spatial Cognition III: Routes and Navigation, Human Memory and Learning, Spatial Representation and Spatial Learning.* Heidelberg: Springer Verlag. 2003

Galton, A. *Qualitative Spatial Change,* Oxford: Oxford University Press. 2000

Giddens, A. *Positivism and Sociology.* London: Heinemann 1974

Gimblett, R.H. ed. *Integrating Geographic Information Systems and Agent-based Modeling Techniques for Simulating Social and Ecological Processes.* Santa Fe Institute, Studies in the Sciences of Complexity. Oxford: Oxford University Press.2002

Goodchild, M. F. "Geographical Information Science". *International Journal of Geographical Information Systems,* 6, (1992): 31-46.

Habermas, J. *The Theory of Communicative Action.* Vol. I: *Reason and the Rationalization of Society*, T. McCarthy (trans.). Boston: Beacon. 1984

Harrington, J.P. 1916. *The Ethnogeography of the Tewa Indians*, twenty-ninth annual report of the Bureau of American Ethnology, Washington, Government Printing Office

Hester Jr. T L. "Choctaw Conceptions of the Excellence of the Self, with Implications for Education". In: *American Indian Thought,* A.Waters, (Ed.) Malden, Blackwell Publishing. 2004: pp 182-187

Hunn, E. "Columbia Plateau Indian Placenames: What Can They Teach Us". *Anthropology* 6 (1996): 3-26

Jojola, T. "Notes on Identity, Time, Space, and Place", in: A. Waters (ed) *American Indian Thought,* 87-96, , Malden, Blackwell Publishing. 2004

Jordan, P. "Continuity and Change in Different Domains of Culture: An Emerging Approach to Understanding Diversity in Technological Traditions". In: *The Model-Based Archaeology of Socionatural Systems.* T.A. Kohler and S.E. van der Leeuw, (Eds.) (pp 13-39), Santa Fe:School for Advanced Research Press. 2007

Kari, J. "A Preliminary View of Hydronimic Districts in Northern Athabaskan Prehistory" *Names,* 44 (1996): p253-71

Kohler T.A. and van der Leeuw S.E., (Eds.) *The Model-Based Archaeology of Socionatural Systems.* Santa Fe:School for Advanced Research Press. 2007

Kuhn, T.S. *The Structure of Scientific Revolutions*, second edition, Chicago: University of Chicago Press 1970

Kwan, M. and Lee, J. "Geovisualization of Human Activity Patterns Using 3D GIS: A Time Geographic Approach". in *Socially Integrated Social Science*. D. Janelle and M. Goodchild, (Eds) (pp. 48-66). New York, NY, Oxford University Press. 2004

Lakoff, G, and Johnson, M. *Metaphors We Live By* . Chicago: University of Chicago Press. 1980

Levinson, S. C. *Space in Language and Cognition, explorations in cognitive diversity,* Cambridge, Cambridge University Press. 2003

Levinson, S. C. "Landscape, seascape and the ontology of places on Rossel Island, Papua New Guinea," *Language Sciences,* 30, (2008): 256-290.

Levinson, S. C., and Wilkins, D (eds). *Grammars of Space,* Cambridge, Cambridge University Press. 2006

Majid, A., Enfield, N.J., van Staden, M. (eds.) Cross-Linguistic Categorization of the Body, *Language Sciences* 28. 2006

Manson, Steven M., D. O'Sullivan, "Complexity theory in the study of space and place" *Environment and Planning A*, 38(4) .(2006):677-692

Mark. D.M. and Frank, A.U. (editors) *Cognitive and Linguistic Aspects of Geographic Space*, Dordrecht: Kluwer Academic Publishers. 1991

Mark, D. M., Turk, A. G., and Stea, D. "Progress on the Yindjibarndi Ethnophysiography", in *COSIT 2007,* Heidelberg, Springer. 2007

Mithun, M. *The Languages of Native North America,* Cambridge, Cambridge University Press. 1999

Nesbitt, R. *The Geography of Thought: How Asians and Westerners Think Differently..and Why*, NewYork: The Free Press. 2003

Ni, F, Cavazo, T., Hughes, M. K., Comrie, A. C., and Funhouser, G. "Cool-season precipitation in the southwestern USA since AD 1000; a comparison of linear and non-linear techniques for reconstruction,"*International Journal of Climate,* 22, (13), (2002): 1645-1662.

Ortiz, A. *The Tewa World, Space, Time, Being, & Becoming in a Pueblo Society,* Chicago, University of Chicago Press. 1969

Peuquet, D. J. *Representations of Space and Time,* new York, Guilford Press. 2002

Raper, J. *Multidimensional Geographic Information Science,* London, Taylor and Francis. 2000

Regier, T. *The Human Semantic Potential: spatial language and constrained connectionism.* Cambridge: MIT Press. 1996

Root, M. *Philosophy of Social Science: the methods, ideals, and politics of social inquiry*. Oxford: Blackwell 1993

Schuurman, N. "Reconciling Social Constructivism and Realism in GIS", *ACME*, 1, (2002): 73-90

Smith, B. "Mereotopology: A Theory of Parts and Boundaries", *Data and Knowledge Engineering*, 20 (1996), 287-303

Smith, L. Tuhiwai, *Decolonizing Methodologies, Research and Indigenous Peoples,* London, New York, Zed Books,Ltd. 1999

SEP, Stanford Encyclopedia of Philosphy, accessed October 14, 2009: http://plato.stanford.edu/

Swentzell, R. "An Understated Sacredness," in: V. B. Price, and B. H. Morrow, *Anasazi Architecture and American Design,* Albuquerque, UNM Press. 1997

Swentzell, R. "Pueblo Space, Form, and Mythology" in: *Pueblo Style and Regional Archtiecture,* N.C. Markovich, W.F.E. Preiser, and F.G. Sturm, editors, New York: Van Nostrand Reinhold. 1990

Wallerstein, I, *The Uncertainties of Knowledge,* Philadelphia, Temple University Press. 2004

Waters, A, (Ed.) *American Indian Thought,* Malden, Blackwell Publishing 2004

Waters, A. Ontology of Identity and Interstitial Being. In: *American Indian Thought,* A.Waters, (Ed.) Malden, Blackwell Publishing. 2004: pp 153-169

Whitehead, A.N. *The Concept of Nature*, The Tarner Lectures Delivered in Trinity College, November 1919, Project Gutenberg: accessed October 14,2009: http://www.gutenberg.org/etext/18835

Whitt, L.A.Biocolonialism and the Commodification of Knowledge. In: *American Indian Thought,* A.Waters, (Ed.) Malden, Blackwell Publishing. 2004: pp 188-213

Williams, J.S. *A means for Oral Cartography from the Use of Metaphor and Metonymy within Navajo Placenames*. MA Thesis, Albuquerque, University of New Mexico. 2000

Zubrow, E. B.W. "Digital Archaeology: a historical context" in: *Digital Archaeology, bridging method and theory,* T. L Evans. and P. Daly, (eds) New York: Routledge. 2006

Wayfinding across space, time, and society

Erik Champion[1]
[1]Auckland School of Design, Massey University.

Abstract

When we talk of cyber-archaeology, it is possible to think of digital archaeology, archaeological methods that incorporate digital technology. Yet this term is too generic to convey the recurring issue of communicating conflicting accounts of the past to a wide variety of audience. We require a term that inspires the participant to move past the contemplation of artefacts as singular objects, and to conserve the situated processes and ways in which this knowledge has been retrieved. Such way-finding necessitates meaningful participation for the visitor to truly engage with the experience, not the superficial personalization or widespread destruction of commercial game-worlds.

***Key words**: virtual heritage, navigation, cyberarchaeology*

1. Virtual Heritage

When archaeology is used to communicate artefacts and events of historical significance to the wider public, we may also be reminded of virtual heritage. Despite its succinct nature, this term is, unfortunately, an oxymoron. Are we attempting to almost convey the importance of preservation and conservation?

In the last few years there have been calls for digital cultural heritage or virtual heritage by Graphite, VSMM, New Heritage Forum, VRST, VAST, DIME, Archäologie & Computer, and DACH, just to name a few. Many of these conference presentations argue that more accurate technology is required, few argue for more understanding of how people use and learn from virtual heritage projects.

In the past I have suggested features for virtual heritage environments that need investigating. These include the issue of cultural presence, hermeneutic richness, learning through interaction, experientially bound worlds, augmented information, shareholder ownership, and non-intrusive evaluation. However the topic discussed here is more immediately pragmatic: how central navigation is to a well-founded theory and practice of cyber archaeology, and the interactive possibilities that new media affords to virtual heritage in this new cyber-age.

Cyber-archaeology is arguably too generic, there should be a term to inspire the participant to move past the contemplation of artefacts as singular objects, and to conserve the situated processes and ways in which this knowledge has been retrieved; a complicit capacity of digital media to allow the visitor to be immersed inside archaeological findings and approaches, way-finding their way through alternating / conflicting viewpoints.

I proffer this definition: cyber-archaeology is the complicit capacity of digital media to allow the visitor to be immersed inside archaeological findings and approaches, way-finding their way through alternating and even conflicting viewpoints.

The hugely successful market of computer games has shown us that unlike traditional media, interactive virtual environments can be highly useful for procedural learning. Traditionally. we typically learn about history, about events, facts, and other cultural perspectives, through prescriptive learning (fig. 1). However procedural learning is more in tune with the recreation of current generations, and allows them to learn by trial and error and with different learning strategies and preferences.

Figure 1: A NWN game mod designed to teach history or literature ("Journey to the West.")

To bring life back to virtual worlds for the enjoyment of the spectator, we require more understanding of communication design, specifically, interaction design. This requires meaningful participation for visitor to truly engage with the experience, not the superficial personalization or widespread destruction of commercial game-worlds or using games that are reflection-friendly and not just trigger-happy.

Game-like environments are seldom taken seriously, students & users are stuck in 'game mode', and there is the issue of authenticity versus interaction: the more interactive the environment the more likely it could be changed in a way not commensurate with history or cultural norms.

From a technological point of view, games can be hijacked, frequently contain serious bugs, and offer only simplistic interaction. They can also disappear, and content can be difficult to port to other, better solutions. Also, meaningfully transferable knowledge and skills are not always obvious.

Professor Susan Greenfield has also argued that the immediacy and short-term excitement of screen interaction is stripping away our ability to follow a narrative and to understand context.[1] Yet this is not a new phenomenon. Vaughan Bell recently uncovered an article about the 19th century neurologists George Beard and Silas Weir Mitchell who were worried about the pace of life and the harmful effect new technologies were having on the brains of American citizens.[2]

Despite misgivings over entertainment media used for educational purposes, we can see various ways in which games and game engines can be used to augment historical and heritage-based learning. Participants can learn via resource management style-games, or online worlds can teach them constrained social behaviour (a European project featuring Blaxxun and entitled *Renaissance* had exactly this purpose).

Travellers and tourists learn about places by going there, observing events, and being instructed by signs, guides and printed material. Their objective is to learn about culture through interaction (otherwise they could read a book), and hence a virtual tourist or traveller would be aided by devices that augment their learning experience via interaction.

What can one do? What can one learn? How does one separate the significant from the accidental and the authentic from the spurious? To extrapolate from a paper by Paul Dourish,[3] in this situation we must grapple with issues of interaction rather than representation, for cultural learning is a highly interactive process.

So I am arguing, in a Heideggeran sense, to deliberately transform the tools of gamic environments from being ready-to-hand, to being present-at-hand. For the tools and interface of a virtual environment or digital environment present a potentially situated cultural viewpoint.

Despite being instruments, they are not objective, and could be realigned to reflect a particularly culturally constrained way of moving, viewing, and acting. Reflecting on specific tools in order to acquire them as ready-to-hand persuades the participant to understand how they are to be used in a culturally performaative context.

With better camerea tracking and improved physical computing interfaces, participants can indulge in chat, learn by observation and mimicry, or select different versions of history. They might be requiired to select the culturally appropriate or historically correct objects or appearance to move about the 'world' or to trade cultural goods, or to advance social roles over a period of time.

Tasks could involve an understanding of local scale, landscape or climate changes, or the need to decipter codices or other linguistic artefacts. Game engines can also afford collaborative storytelling (in-world role-playing and film-making through machinima tools).

Mapping

Maps are often viewed as straightforward devices to locate and orientate one's position and location. Yet they are not just instrumental artefacts but also epistemic ones with a long history: they have helped organise our knowledge of the world for many millennia.

David Modjeska defines navigation via locomotion and wayfinding:

> ...a useful definition of navigation in electronic worlds: locomotion + wayfinding. Locomotion is the activity of moving from one location to another.
> Wayfinding is primarily a cognitive process, comprising three abilities:
> -cognitive mapping or information generation to understand the environment
> -decision making to structure and plan actions
> -decision execution to transform decisions to behavioral actions.[4]

Rather than seeing wayfinding as a process, Elvins seems to see wayfinding as an ability:

> "Wayfinding is "the ability to find a way to a particular location in an expedient manner and to recognize the destination when reached", i.e. navigating with spatial knowledge and a destination in mind. [5]

However, both Modjeska and Elvins argue that maps in virtual environments should not serve merely as abstractions of reality. They both argue that navigation will be more usable if maps are similar to the way our spatial memory is created recalled and acted upon as part of a decision making process.

Navigation is not a simple activity, it involves situating oneself in a world, creating an overview of the locality (often involving specific tools or techniques), and

1 Andrew Marr, (interview with Susan Goldsmith), radio interview, *BBC radio 4*, 19 May 2008. http://www.bbc.co.uk/radio4/factual/starttheweek_20080519.shtml

2 Vaughan Bell, "Is the cinematograph making us stupid?" Mindhacks blog, 28 June 2008. http://www.mindhacks.com/blog/2008/07/is_the_cinematograph.html

3 Paul Dourish, *Where the Action Is : The Foundations of Embodied Interaction. (*Cambridge, Mass.: MIT Press, 2001).

4 David Modjeska. *"Navigation in Electronic Worlds: Research Review for Depth Oral Exam".* Department of Computer Science. 10 January 2005 <http://www.fis.utoronto.ca/faculty/modjeska/Pubs/lit_rvw.pdf

5 T. Todd Elvins. "Virtually lost in virtual worlds -- wayfinding without a cognitive map". *ACM Computer Graphics.* 31.3 (1997): 15-17.

organizing goals, then calculating (often in a procedural rather than preset way) how to reach these goals.

In his online paper on interactive maps, "Contextualising Archaeological Information through Interactive Maps",[6] Ian Johnson lists many possible uses of digital maps, for example, that they can act as cultural representations. However, in his case studies of interactive maps, he concentrates exclusively on the ability of digital maps to interactively access and filter database-driven data.

While this may be important to scientific applications that require immediate access to large amounts of frequently changing data, there are many other interactive possibilities of digital maps still to be explored as innovative and potentially fruitful research areas.

Particularly in terms of cyber-archaeology, navigaton is also more than physical activity. One must learn how to understand (overview) and move between landmarks, changes in time scale, and differeing points of view (both socially, and in terms of what was recreated or otherwise envisaged).

So there are many hybrid possibilities for digital maps used in conjunction with other digital media, such as panoramas, and three-dimensional virtual environments. Further, a digital map as a cognitive artefact, a cultural representation, a social medium, and as a potential evaluation tool. Research into such areas may help not just conventional navigation and wayfinding, but also the social interaction of participants, and their cultural understanding of real and virtual environments.

Cognitive Maps

There is confusion in the literature over whether a cognitive map is the process by which people store navigational knowledge or whether it is the instantaneous product of cognitive mapping, i.e. a cognitive map is formed in the mind on demand from cognitive mapping processes.

Lagoudakis for example, seems to define a cognitive map as the former; but Soini defines it as the latter, 'a product of this [cognitive mapping] process at any point in time'.[7] Medical research indicates that Soini is more likely to be correct: cognitive maps are created on demand as their elements are retrieved from different parts of the brain on demand. In fact, research (Wang Johnson and Zhang 2001, 191-98) suggests that 'space is represented in the mind not once but multiple times'.[8]

Soini also distinguishes between the term 'mental map', as those maps that people draw when they are asked to sketch out their cognitive maps, and concept maps. She defines a concept map as 'a graphic system for understanding the relationship between concepts'.

Soini also defines symbol mapping, as freehand mapping of places visited by selecting symbols that represent different personal meanings, such as 'nice', 'beautiful', 'private' etc. She then suggests there may be many interesting 'hybrid' ways of combining (and evaluating) these different types of mapping.

Given her thorough investigation of the terms, I will follow Soini's definitions of mapping, and refer to a person's cognitive navigational knowledge as cognitive mapping, the image or schematic they process from their cognitive mapping processes on the fly as a cognitive map, and their drawn interpretation of their cognitive map as a mental map. Mental maps could include sketch maps, symbol maps, iconic maps (symbols with no specific personal connotations, or concept maps.

I stress that cognitive maps are not the same as maps created by professional cartographers. Neural research by scientists on monkeys indicates that we remember locations in terms of salience (behavioural significance) not by what is actually there.[9] Thus, the way we access these cognitive maps is typically not just via quantitative estimates and measurements but also in relation to personal attachments and perceptions. For example, Bishop and Green have written:

"The Australian landscape is shaped by the values of the inhabitants'...people's values change with their changing perceptions and understanding of the landscape."[10]

Raubal and Egenhofer[11] summarise the wayfinding literature and emphasise the distinction between 'practical space (i.e., acting in space) and conceptual space (i.e., representing space)'.[12] They refer to Mark Johnson's belief that wayfinding is developed by habits and social practices, along with image schemata, 'recurring mental patterns that help people to structure space so that they know what to do with it'.[13]

Maps And Culture

Cultural learning could be summarised as learning through observation, instruction, or by trial and error. Therefore, there are two major ways of transmitting culture: through other social agents (through the language actions and

6 Ian Johnson, "Contextualising Archaeological Information through Interactive Maps." *Internet Archaeology* 12 (2002). 20 May 2005 <intarch.ac.uk/journal/issue12/johnson_index.html>.

7 Katriina Soini. "Exploring human dimensions of multifunctional landscapes through mapping and map-making." *Landscape and Urban Planning* 57.3-4 (2001): 225-239, p. 229.

8Hongbin Wang, Todd Johnson and, Jiajie Zhang. "The mind's views of space." Proceedings of the 3rd International Conference of Cognitive Science. Beijing: China (2001): 191-198.

9 Jacqueline P. Gottlieb, Makoto Kusunoki and Michael E. Goldberg. "The representation of visual salience in monkey parietal cortex." *Nature* 391.6666 (January 29 1998): 481-484.

10 Ian Bishop and Ray Green. *Community Exploration of Changing Landscape Values.* Project Proposal. Funded by: Land and Water Resources Research and Development Corporation. 12 Dec. 2004 <http://www.abp.unimelb.edu.au/Research/Ind/CLV.html>. Unpaginated.

11 Martin Raubal and Max Egenhofer. "Comparing the Complexity of Wayfinding Tasks in Built Environments." *Environment and Planning B*, 25.6 (1998): 895.

12 Ibid.

13 Mark Johnson. *The body in the mind: The bodily basis of meaning, imagination, and reason.* (Chicago: University of Chicago Press, 1987).

reactions of other people), and through artefacts, (the objects created and modified by people). The former seems necessary for understanding a culture natively (from the inside as vicarious experience), and the latter seems necessary for extending cultural knowledge or developing cultural awareness of alterity (from the outside as observation or as extrapolated experience).

Cultural presence exists via artefact-mediated communication between social agents (the creator and the recipient), even when the creator is not there. The ability of an artefact to convey a sense of that creator's agency is a reflection of its 'hermeneutic richness' (akin to the archaeological notion of the 'trace'). The perceived sense of that creator's agency through an artefacts is itself cultural agency. For an artefact is itself a cipher, a mark of cultural agency.

Kim Sterelny[14] traced the ancestors of the map (cave paintings) back over 30 000 years, 'With the invention and elaboration of pictorial representation, humans came to be makers of specialised epistemic artefacts'. As epistemic artefacts ('tools for thinking'), that is, items that structure our knowledge outside of our minds, maps are '"portable" representational resources'. Even though maps are abstract, they are still part of the cultural process; we must learn to read them for the rules of abstraction are derived from cultural rules.

Western notions of reality (of what is out there) are thus only part of the features of cognitive mappings for other cultures. There are other less straightforward and more social and cultural aspects to maps.

One example of a cognitive map as a socially specific and culturally constrained artefact is the Aboriginal painting. The text in the article *Aboriginal art online* notes: [15]

Nearly all Aboriginal art can be related to landscape and some paintings and designs do represent explicitly the physical relationship between different features of the landscape. However, Aboriginal paintings should be seen primarily as maps of conceptual relationships that influence the way the landscape is seen and understood. When Aboriginal paintings do represent specific features of landscape, they show them in their mythical rather than their physical relationship to one another.

While modern day mapping professionals view the maps they make as an objective and accurate abstraction of reality, maps are actually culturally specific and socially constrained. They are designed via social conventions for a specific audience, and are abstracted to help people's cognitive mapping.

Interactive Maps

Maps in virtual environments are needed to overcome the limited navigational aids of virtual environments. Disorientation in virtual environments is a frequent and troubling issue for many users, an issue noted by various writers, hence a map plays an important practical aid in navigating virtual environments. For virtual environments - even those with Head Mounted Displays (HMDs) and motion tracking - lack many orientating features of the real world (for a list of these features refer Vinson)[16].

Digital maps also offer other advantages over real world maps. Being interactive, recordable, and dynamically personalizable, digital maps can gradually or contextually reveal the unknown, keep a personally relevant record of the past, or offer a shadowy gate to the future.

A foreteller of what lies ahead, or what is a worthwhile digression or quicker way around a yet to be seen obstacle, real world maps help 'preview' possible and optimal journeys. Yet digital maps used for previewing (wayfinding) can also be directional and proxemic. As users progress through the virtual environment, a digital map could improve in local accuracy. This feature is already used in certain types of games, and could help sustain mystery, but also reduce cognitive loading.

For example, imagine that on entering a virtual environment for the first time, that the map is faded and only shows a very blurry concept of what is in the area, as found on maps by neighbours. As one explores the virtual environment, the map becomes more accurate. Another game theme is to only show a radial area around the player, this creates dramatic tension, stresses the danger of enemies to sneak up from behind, and encourages the player to continually move (and re-orient) through the environment (see fig. 2).

Figure 2: A limited compass map increases dramatic tension (UNSW student project).

14 Kim Sterelny. "Externalism, Epistemic Artefacts and the Extended Mind". Online papers, 12 Dec. 2004 <http://www.vuw.ac.nz/phil/staff/sterelny-papers.aspx>, 2004: 5.

15 Aboriginal art online. *Art, Land and the Dreaming.* Online text, 12 Dec. 2004. <http://www.aboriginalartonline.com/culture/land.html>. Unpaginated.

16 Norman G. Vinson "Design guidelines for landmarks to support navigation in virtual environments." *Proceedings of the SIGCHI conference on Human factors in computing systems: the CHI is the limit.* Pennsylvania, United States, (1997): 278–285.

Digital maps could also be related more closely to a sense of place. Consider a virtual environment that has dynamic place elements that may both afford and constrain progress. Perhaps changes in light could symbolise day and night.

Night may be the metaphorical resting time where a user needs to find a safe place to sleep (recharge). Coupled with an interactive digital map the user could be induced to plan and record (via the map) all paths and journeys so that they do not fall into a stupor in a dangerous place (from which they cannot move or wake up from).

Any device for orientation will help users navigate through an environment but an interactive digital map may further allow a graphical history of their virtual travels. For example, Ramloll and Mowat[17] created a virtual environment where participants could take in-scene snapshots of exit and entry points. They could retrieve these snapshots at any time and click on them to move to those specific locations. The writers reported that this function increased ease of navigation.

On the other hand, Nitsche and Thomas[18] describe a project that uses cinematic devices and teleports assailants to confuse and hinder the participant's mental mapping of an environment. They believe these steps increase dramatic tension and help make small environments appear larger (and hence more interesting for the participant).

Unlike typical conventional maps, digital maps can automatically filter information according to current position or individual preferences or hidden (or targeted) events in a virtual environment. They can also be used like an inventory, to keep a reminder of what options or resources are currently accessible.

In the sense that maps can act as memory aids to wayfinding, they are also 'memory' maps, graphical travel diaries. Digital interactive maps can help orient, help navigate, and recollect past episodes along a journey. For digital interactive maps can 'remember'. Such memory maps could record vistas, landmarks, use of artefacts, and encounters-they can act as a form of pictorial diary. In order to help memory recall users may be allowed to select scale and position thumbnail icons of events, encounters, or artefacts onto their map.

For example, users could customize from a palette of icons (castles houses bridges, food sources etc), select, scale, and position these pictorial icons to remind them where they went, what they did and where the safest routes are etc. Billinghurst and Weghorst[19] have previously evaluated engagement by qualitative and quantitative measurement of 'sketch maps'. Their research indicates that accuracy in sketching the virtual environment after the experience is directly related to engagement in the virtual environment.

Further, as artefacts, the way maps are used, personalised and aged, are 'traces' of their 'life-history'. Maps are epistemic artefacts to aid perception and decision-making, but in being used, they reflect popular and annotated points. Most importantly for us, an interactive personalizable map could show where people went, and how effectively they avoided the areas dangerous to their metaphorical 'health' points in the above game-style scenario. Such annotations would reveal the popularity of certain spots in the virtual environment.

For the cognitive map (as it is called in Human-Computer-Interaction studies and town-planning literature) is a long-accepted model of human wayfinding and hence an appropriate tool to evaluate. While research indicates there is no one primary map the same research does indicate that our cognitive mapping is multimodal and hierarchical. Given this, it seems reasonable to presume that a multimodal graphical mapping device that people indirectly customize (indirectly as in the customisation arises through task selection) will enable the navigation to be more intuitive and hence usable.

Any device for orientation will help users navigate through an environment but a map further allows a graphical history of their virtual travels. If actors are asked to manually select scale and place icons to represent where certain activities or important cues are, then we might be able to track via the size and placement of icons some degree of their engagement.

One type of evaluation is the 'speak aloud' method. By extrapolation, digital maps may help a 'talk aloud' method. A digital interactive map can help users graphically plot the quickest or most scenic journey without entering hostile or physically arduous areas (physical constraints) and when users are asked to lead others through the virtual environment (using the 'teach' evaluation method) the map will help as a visual teaching cue.

For personal orientation and memory recall, i.e. to keep mementos relating to special events, participants can select, scale, and position thumbnail icons of events, encounters, or artefacts onto their map (here known as a 'memento map'). Users could update the memento maps with their own sized positioned and scaled thumbnail icons.

These icons when clicked on will hyperlink to the time and location of the event encounter or landmark recorded. The frequency, accuracy, and sizing of icons will indicate their

17 Rameshsharma Ramloll and Darren Mowat. "Wayfinding in Virtual Environments using an Interactive Spatial Cognitive Map." *The Fifth International Visualisation 2001 conference, IV2001* Proceedings, London: IEEE Press, 2001. <http://www.dcs.gla.ac.uk/~ramesh/Online Publication/ramlollr_wayfinding.pdf>.

18 Michael Nitsche and Maureen Thomas. "Stories in Space: The Concept of the Story Map." *Proceedings of the Second Conference on Virtual Storytelling ICVS '03*, ed. by Olivier Balet, Gerard Subsol, Patrice Torquet, Berlin et al.: Springer Verlag (2003): 85-94.

19 Mark Billinghurst and Suzanne Weghorst. "The use of sketch maps to measure cognitive maps of virtual environments." *VRAIS Proceedings of the Virtual Reality Annual International Symposium (VRAIS'95)*. IEEE Computer Society, Washington, DC, USA (1995): 40.

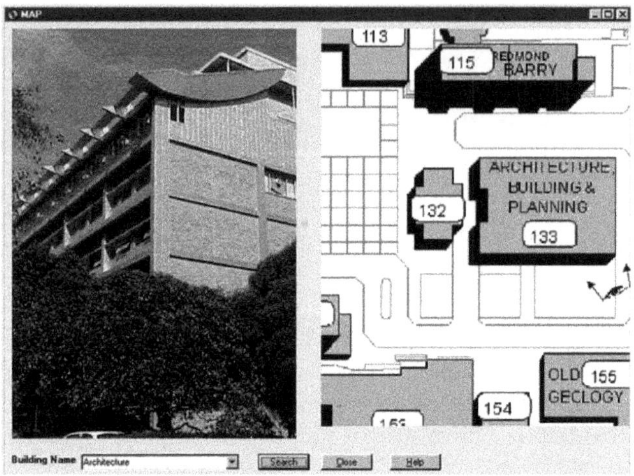

Figure 3: The ShadowFinder-a hybrid 2D-3D map (University of Melbourne student project).

amount of care and concern with the landmarks. When their virtual journeys are over, memento maps could be downloaded to more portable media (for personal later journeys in the real world), or they could be sent to friends and to family around the world so loved ones could track the virtual paths and annotations of real world travellers.

2D and 3D Maps

The above project involved usability testing of a hybrid 2D-3D mapping application, (fig. 3). This project was run in collaboration with three postgraduate students. The aim was to create a hybrid mapping system of 2D plan and 3D panorama that could run on an A5 and A4 sized screen. Participants could navigate around the 2D map (on the right of fig. 3), and their 3D view (to the left of the picture) would automatically update to accommodate the new eye position. They could also search for buildings via an attached database, and successful searches would update the screen with new map positions and 3D views.

GPS could be added so that the 3D view would synchronise with the user's real world-view. That is, when one moves the cursor over a plan, the 3D view automatically updates so that users can see where they are on the plan. The 3D view should rotate, change scale, and update the screen to correspond to what the user sees in front of them in the real world.

The usability testing (of 4 participants) was to see if the prototype was easier and quicker to use than a conventional paper-based map, and to ascertain whether a smaller A5 sized screen would be big enough for adequate task performance. The answer to both questions was in the affirmative.

Event Mapping in 3D worlds

One could also apply this idea to online virtual environments delivered over the Internet (fig. 4). In moving the cursor over a 2D map, the web-based application would automatically change the avatar's location and orientation in the accompanying virtual environment.

Figure 4: Dynamic Map (HSWorlds).

A scripted template still available over the Internet, created the above interactive map, designed with the help of the educational company HSWorlds. It allows designers to select a virtual environment and a map, type in the dimensions of the virtual environment and the map, and the interface will create the necessary script to activate the map so that when clicked it transports visitors to that point in the world. It will also add buttons for strategic points and drop the visitor from a predetermined height (if the terrain is hilly).

PanoMaps

A simpler variant of this hybrid map idea could also be used for real world navigation and orientation but using a panorama instead of a 3D virtual environment. The Panomap is a hybrid 2D map with an accompanying 360-degree panoramic photograph (fig. 5).

Through moving around on the 2D map, the panorama automatically rotates around to match with the orientation of the map view. Information could be triggered from either spinning around in the panorama or moving the cursor or rotating the cursor on the 2D map.

One could even have an infinite roll of panorama that if one spins around in more than 360-degrees, the panorama changes to a digitally augmented panorama of

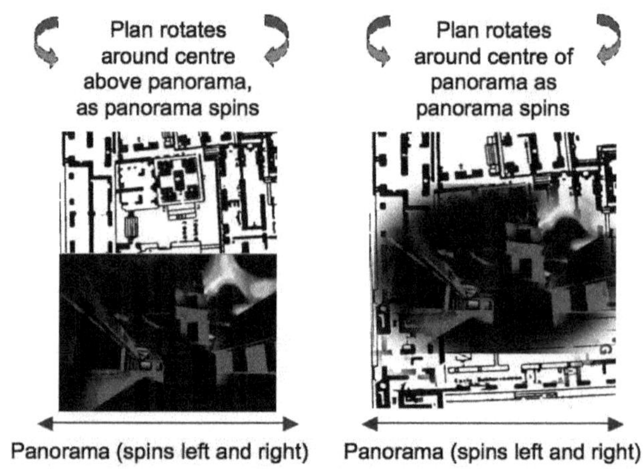

Figure 5: Two examples of a Panomap.

Figure 6: Siena Piazza del Campo

what used to be there (that is, changes in view orientation can change not just the orientation in the 3D environment, but also the chronological data in the 3D environment).

I believe the 2D map with popup panoramas may be particularly good for older towns and villages, where the panoramas correspond to plazas (fig. 6). In small medieval towns, the major circulation paths often converge to a major town square or piazza. These paths are often crooked, cross each other regularly, or vary in span, and are not easily identifiable by recourse to a traditional 2D map.

For these types of plaza-designed urban environments, the Panomap idea may be more efficient and user-friendly. One simply selects a suitable sized plaza on the 2D map, and the related panorama of a plaza opens for the user to identify.

Once the user determines that is the correct piazza, they can identify their orientation and where the major related landmarks and routes are, by spinning the panorama, which in turn rotates the accompanying plan view. The common problem of knowing which way is 'up' on the map is thus no longer an issue.

For even more convenience, a combined GPS and a compass on a portable device (such as a PDA) could automatically locate and position the viewer via the 2D map. The screen could be split into this map view, and a panorama photograph that automatically rotates to which parts of the piazza the user is facing (fig. 4). Hotlinks could suggest important pathways or landmarks or the orientation of nearby piazzas.

There is also a practical reason to use a Panomap visualisation instead of the hybrid system of 2D map and 3D virtual environment. If we have a suitable panoramic mapping technique proves effective, it means we would not have to model every building in a town in order to facilitate portable travel information and travel experience.

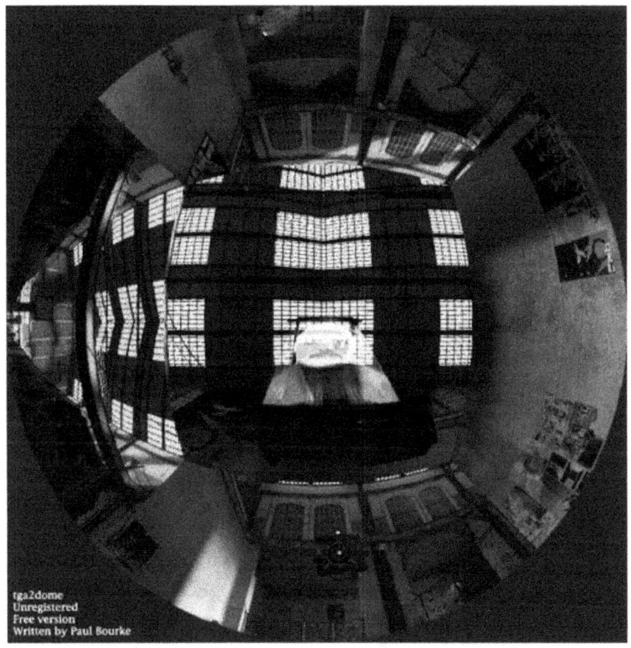

Figure 7: Many games can automatically create panoramas (Half-Life 2 / Source game engine)

Current game engines can automatically create high resolution panoramas (fig. 7). They typically feature panoramas as skyboxes but they can sometimes integrate realistic panorama bubbles inside virtual environments, fading in or fading out as the participant arrives or leaves a 'hot spot'.

Attitudinal MiniMaps

Various virtual environments have used ideas like world maps or minimaps, small-scale pictures within pictures of the world, to allow quick orientation through large virtual environments. Generally, these ideas are to facilitate navigation and orientation, but they could also be used thematically to increase immersion in virtual environments, show the behavioral states of other players, or reveal different perspectives of the same 'world'.

In virtual environments, users do not have access to peripheral visual information available to them in reality (due to the lesser field of vision, no sideways glances and no natural sideways motion of the eyes through ambulation). In game design, the player often wishes to know where enemies are without having to leave the screen. Designers often create a hybrid visual environment of 3D world and translucent overlay map as a way of getting around these problems. Rather than directly recreate the 3D environment as plan objects, game designers often represent the moving players and rotate the map view to always reflect what is in front of the player.

For example, in the game Lugaru (fig. 8), the overlay map rotates with orientation of the player's bunny avatar, and reveals only the plan outlines of buildings.

Lugaru not only shows other moving characters and

Figure 8: Lugaru-an example of a game using an overlayed rotating map (lower right corner).

where they are facing (the enemies are represented by triangles), but the moving map symbols of the enemies change colour to indicate if they are just walking, in a sensing state, or have found and are chasing the player's avatar.

Camera Window Maps

Clive Jackson of Blink3D has created a programming API that allows the screen of a web-based virtual environment to be overlaid by smaller screens. In Blink3D this technique is called picture within picture (fig. 9), and it may be used to create a second camera view as an overlay map or secondary perspective view inserted into the main screen of the virtual environment.

More than just a camera view masquerading as a mask, the secondary camera window's features can be changed programmatically (thematically), dependent on player interaction or environmental change. As it can be rescripted, the camera window can change in field of view, from architectural to perspective, in position and size, or what it is attached to or focused on, and be dependent on what happens in the main virtual window. There could be more than one secondary camera window, and the player may be able to interact between the various camera windows as *different perspectives* on the same world.

I have suggested that the interactive possibilities of hybrid digital maps are by no means fully exhausted, for we have not fully examined how we use maps in the real world, and how maps can incorporate digital media. For example, the paper map you have at home may no longer be the same as the map you originally bought. Now your map could be more like a palimpsest, it may now have scrawled over it your annotations, diagrams, and instructions to self. The map's condition will also reflect the way you use and care for maps.

Digital maps can also be tailored to user needs rather than to raw data, personalised by the user or by diverse and varying audience needs, and made more meaningful by directly indicating the goals, tasks and achievements of the user. Maps can also be hybrid in another sense,

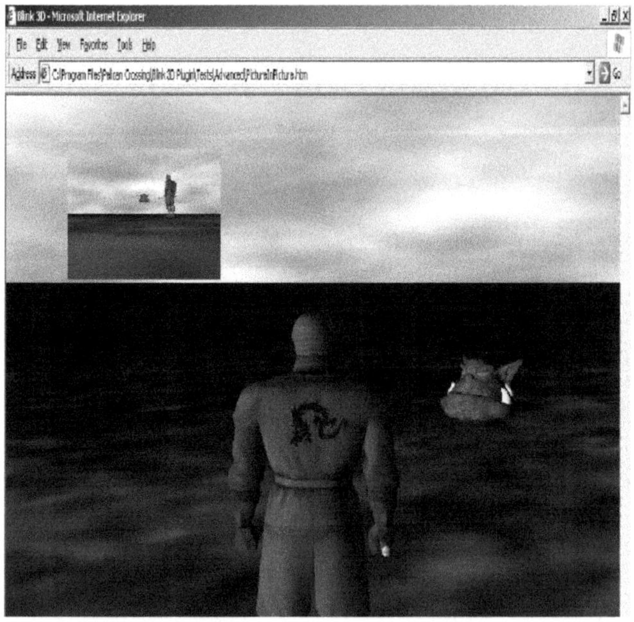

Figure 9: 'PictureInPicture' screenshot of Blink3D showing secondary camera window.

through combining the multimodal features of digital media. When incorporated into a 3D environment as a hybrid mapping and environment visualisation system, such maps can interact with the different physics and laws of virtual environments, filter significant data, auto-record interaction use, and augment the collaborative potential of multi-user virtual environments.

Through recording user interaction and preferences, they may function as passive evaluation tools. Where they can be annotated or otherwise modified by participants, this personalisation may also be shared amongst participants and non-participants.

Through such emphasis on user interaction and user personalisation of key information, personal goals and achievements, they may also help enrich social participation in virtual environments. Digital maps are thus more than simple database interfaces. When they leverage the interactive potential of digital media, maps can also regain their historical usage as cultural artefacts, cognitive feedback mechanisms, and epistemic tools.

Navigation Prototypes

There are of course wider navigational implications than the use of game-based mapping themes for virtual environments. The artistic expansion of this digital entertainment explosion in terms of enhanced sensory input and output may also help virtual reality break free of the mouse and the screen in order to thematically and appropriately transmit the significance of past cultures and distant societies.

Peripheral Immersion

Thanks to various software and hardware innovations such as CAVE UT (fig. 10), or Paul Bourke's warping algorithims,

Figure 10: Surround Projection (CAVE UT).

Figure 11: Mayan temple city.

we can know insert panoramas into virtual environments or cheaply create curved screen projections. The upshot of these largescale and more display technologies is that our peripheral vision adds to the sense of immersion, and there is now added incentive to tailor virtual worlds to the peripheral and territorial psychological triggers of the paprticipants. This in turn may heighten the sense of a physical world, and the sense of embodiment.

However there are issues still to be tackled. How can we contextualise the interface devices and screen GUI to the culture portrayed? In the case of hybrid panorama environments, will the GUI elements be easy to opearte without being obtrusive? Is spinning around inside a panoramic bubble intuitive or possibly inducing vertigo and confusion?

Here I will briefly outline two examples of the importance of peripheral vision, both are from game-related projects, but both could be employed for cyber-archaeology.

In the first project the University of Queensland (UQ) student team was asked to use the limitations of a head mounted display (HMD) in a creative manner. The HMD offered reasonable resolution but was annoyingly slow in updating the picture. So we decided to use this issue as an atmospherically useful element.

The scenario was an imaginative recreation of the story of the Minotaur and the (Cretan) labyrinth. The player would wear the HMD, hold a plastic sword (which when moved would move a virtual sword), and was asked to hunt down the Minotaur. In this scenario the navigation was coupled with biofeedback gained from previous players to heighten the sense of drama and historic inevitability.

As they reached the end of a tunnel in the virtual maze, one of four fans surrounding the player would turn and blow from the direction of the virtual wind. As the player moved closer to the minotaur the sound of their heartbeat as well as the sound of the heartbeat of previous players would be amplified. Adding to these atmospheric elements, the slowness of the HMD would increase the sense of fear and the slow ability of vision to adapt to darkness.

Figure 12: the Mayan underworld.

The second project required the player to traverse a Mayan temple site (fig. 11). We ported the models from the author's virtual environment to a commercial game engine, UT2004. The compass was derived from a Mayan calendar, and could be used to teach the player Mayan glyphs.

The player had to undertake rituals and tasks in the archaeological environment in order to descend to xibalba, (an artistic recreeation of the Mayan underworld – see fig. 12). The inbuilt navigation tools of the UT2004 game were employed to aid in navigation for such a large virtual environment.

The final environment was shown inside a box, using a 3D joystick, an arduino pad that one physically walked on to virtually walk, and a curved mirror inspired by Paul Bourke's warping research which allowed him to project virtual environments onto curved surfaces (such as the planetarium domes), creating a 3D effect (fig. 13).

Map as Inventory

In another UQ student game design project, the students devised the map as an interactive inventory (fig. 14). In order to gain poweer in the game (the powers of the

Figure 13: Prototype of temple using curved mirror.

Figure 14: Hieroglyph collection game

ancient Egyptian gods), the player had to find and store the appropriate hieroglyphs. Being able to recalll the correct hieroglyphs allowed them to cast godlike powers to save themselves from the skeleal guardians of the temple.

Despite its cartoon-like Indiana Jones theme, this game has some thematic promise. It has a simple to understand interaction metaphor, progress relates to memory recall of hieroglyphs and navigating through and deciphering the hieroglyphs is a vital part of the gameplay.

Navigation by writing

Unlike many games, virtual heritage environments also have a set narrative to tell. How do we allow freedom of interaction and personalization along with the unveiling of history through one or more narratives? Can we infuse written history with multiple personal and cultural perspectives?

In another UQ student project, the students created a game that required the player to trace over Chinese calligraphy using a light tablet (fig. 15). While this does

Figure 15: Chinese Calligraphy Tracing Game

not initially appear to be a navigation related tool it is interesting for several reasons. Interaction can involve the student learning different writing forms, interaction is creative but not destructive, and the player could potentially leave handwritten messages for other players in the virtual environment. The script automatically judged if the player's tracing was accurate enough before it allowed the game to proceed.

Embodied Navigation and Traces

There are other ways to personalise and annotate virtual worlds. In an honours project, Andrew Dekker created a biofeedback game whereby the biofeedback of individual players affected the cinematics (shaders, filters, screen shake), gameplay (such as seeing through walls), and musical atmosphere as well as the artifical intelligence of the non playing characters.

Figure 16: Biofeedback affects display, AI, music.

The screen literally changed according to the stress (galvanic skin response) of the player (fig. 16). It would not be difficult to augment the virutal environment with the biofeedback traces of previous players. These traces could be seen as tunings or auras that reflect their stress, task performance or personalised sites or events.

Diurnal Navigation

We now have accessible technology that can create realistic lighting, as well as real-time climatic effects and hence archaeolgical events and ceremonial ocassions (fig. 17). For example the Mayan culture had a fascinating if terifying view of the difference between night and day, spirit avatars would roam the night and shamans were sent to retrieve cursed souls.

The Mayans also fashioned a detailed astronomical record that led to them creating observatories that replayed specific cosmic events at specific times. It would be feasible to create different navigation styles dependend on whether a virtual Mayan world was in day or night, or if it recreated a specific astronomical event (such as the rise of Venus).

Figure 17: Mayan night and day are different worlds (Dylan Nagel, Paladin Studios).

Conclusion

Interaction is crucial in the creation of culture, and, by extension, in the understanding of culture. Where our environment refers to a long extinct civilisation, such a bridging is perhaps impossible, unless we somehow can bring the ghosts of the culture back to life. In other words, a feeling of strong cultural presence requires being physically *embodied (*we have a body that affects and is affected by other objects and forces), socially *embedded (*there exists the presence of others to whom we feel socially bound) and culturally *inscribed* in the world (our actions leave a lasting and meaningful impression on the world).

So the issue of navigation and interface tools in general is of critical importance to the subjective sense of cultural immersion. Adding the tried and tested navigation techniques of game environments may help cognitive processing (although they can also be deliberately employed to thwart cognitive processing), but they may interfere with the cultiural learning goals of cyber-archaeology.

In a previous evaluation that compared subjective user responses to the task performance and personal ranking in virtual environment projects, I noticed that gamers concentrated on solving tasks rather than exploring the environment for its own sake. And secondly, the more authentic environments were actually considered less authentic the more they afforded interaction.

Hence, to simply add interaction to a virtual environment does not necessarily make it more authentic, and adding game-style scenarios seems to help users know what to expect, but they seem to take it less seriously the more they can affect it. The navigation aids were not culturally appropriate, and constrained in such a way that participants move from ready-to-hand to present-at-hand.

The navigational tools were not thematically and cognitively part of the cultural learning environment. We require a more refined understanding of navigation as a process that allows people to understand and interact with geophysical forms, chronological rifts and attitudinal divides. The etymological origin of *cyber* (Greek for steersman) is of great importance for cyber-archaeology.

Far too many virtual heritage environments do not *yet* address the cognitive limitations of flat-screen virtual worlds, nor do they provide suitable tools to help different people with different learning preferences to locate and wayfind cultural information across space, time, and society.

Acknowledgements

I would like to thank students and staff at Interaction Design, ITEE, University of Queensland, fellow students at the University of Melbourne for the shadowfinder project, and a student game design group at the University of New South Wales, Dr Jeffrey Jacobson at publicvr.org, Stuart Forbes at HSWorlds, and Dylan Nagel at Palidin Studios.

Bibliography

Aboriginal art online. *Art, Land and the Dreaming*. Online text, 12 Dec. 2004. <http://www.aboriginalartonline.com/culture/land.html>.

Billinghurst, Mark, and Weghorst, Suzanne. "The use of sketch maps to measure cognitive maps of virtual environments." *VRAIS Proceedings of the Virtual Reality Annual International Symposium (VRAIS'95)*. IEEE Computer Society, Washington, DC, USA (1995): 40.

Bishop, Ian, and Green, Ray. *Community Exploration of Changing Landscape Values*. Project Proposal. Funded by: Land and Water Resources Research and Development Corporation. 12 Dec. 2004 <http://www.abp.unimelb.edu.au/Research/Ind/CLV.html>.

Darken, Rudolph, and Silbert, John L. "Navigating Large Virtual Spaces". *International Journal of Human-Computer Interaction* 8.1 (January-March 1996): 49-72.

Dourish, Paul. *Where the Action Is: The Foundations of*

Embodied Interaction. Cambridge, Mass.: MIT Press, 2001.

Elvins, T. Todd. "Virtually lost in virtual worlds -- wayfinding without a cognitive map". *ACM Computer Graphics*. 31.3 (1997): 15-17.

Gottlieb, Jacqueline P., Kusunoki, Makoto, and Goldberg, Michael E. "The representation of visual salience in monkey parietal cortex." *Nature* 391.6666 (January 29 1998): 481-484.

Johnson, Ian. "Contextualising Archaeological Information through Interactive Maps." *Internet Archaeology* 12 (2002). 20 May 2005 <intarch.ac.uk/journal/issue12/johnson_index.html>.

Johnson, Mark. *The body in the mind: The bodily basis of meaning, imagination, and reason*. Chicago: University of Chicago Press, 1987.

Lagoudakis, Michail G. "Spatial Knowledge in Humans, Animals and Robots". 12 Dec. 2004 <http://www.cs.duke.edu/~mgl/papers.html>.

Marr, Andrew. Andrew Marr, (interview with Susan Goldsmith), radio interview, *BBC radio 4*, 19 May 2008. <http://www.bbc.co.uk/radio4/factual/starttheweek_20080519.shtml>.

Modjeska, D. *Navigation in Electronic Worlds: Research Review for Depth Oral Exam*. Department of Computer Science. 10 January 2005 <http://www.fis.utoronto.ca/faculty/modjeska/Pubs/lit_rvw.pdf> or <http://citeseer.ist.psu.edu/modjeska97navigation.html>.

Nitsche, Michael, and Thomas, Maureen. "Stories in Space: The Concept of the Story Map." *Proceedings of the Second Conference on Virtual Storytelling ICVS '03,* ed. by Olivier Balet, Gerard Subsol, Patrice Torquet, Berlin et al.: Springer Verlag (2003): 85-94.

Ramloll, Rameshsharma, and Mowat, Darren. "Wayfinding in Virtual Environments using an Interactive Spatial Cognitive Map." *The Fifth International Visualisation 2001 conference, IV2001 Proceedings*, London: IEEE Press, 2001. <http://www.dcs.gla.ac.uk/~ramesh/Online Publication/ramlollr_wayfinding.pdf>.

Raubal, Martin, and Egenhofer, Max. "Comparing the Complexity of Wayfinding Tasks in Built Environments." *Environment and Planning B*, 25.6 (1998): 895-913.

Soini, Katriina. "Exploring human dimensions of multifunctional landscapes through mapping and map-making." *Landscape and Urban Planning* 57.3-4 (2001): 225-239, p. 229.

Sterelny, Kim. "Externalism, Epistemic Artefacts and the Extended Mind". To appear in: Richard Schantz (ed) *The Externalist Challenge. New Studies on Cognition and Intentionality*. Berlin & New York: de Gruyter, 2003, p.5. Online papers, 12 Dec. 2004 <http://www.vuw.ac.nz/phil/staff/sterelny-papers.aspx>.

Vinson, Norman G. "Design guidelines for landmarks to support navigation in virtual environments." *Proceedings of the SIGCHI conference on Human factors in computing systems: the CHI is the limit* Pennsylvania, United States, (1997): 278–285.

Wang, Hongbin, Johnson, Todd, and Zhang, Jiajie. "The mind's views of space." *Proceedings of the 3rd International Conference of Cognitive Science*. Beijing: China (2001): 191-198.

3D Cybermaps of Western Han Mural Tombs

Fabrizio Galeazzi, Paola Di Giuseppantonio Di Franco and Nicolo Dell'Unto
University of California, Merced, USA.

Abstract

The Virtual Museum of the *Western Han Dynasty Project* is a joint research between UC Merced and the Jiaotong University aimed to the digital documentation of archaeological sites, artifacts and cultural relics of the Western Han Dynasty. The outcome of this process will be the creation of a virtual museum, based on collaborative environments, dedicated to the Western Han Dynasty and able to integrate new archaeological datasets coming from the fieldwork activities (most part of them unpublished), monuments, and famous collections of artifacts of the Xi'an archaeological museums.

One of the most important archaeological examples in Xi'an is represented by the mural paintings of the monumental tombs of the Western Han Dynasty. Despite their cultural and historical importance they are at risk of being lost because of the critical conditions of plasters and colors. The murals show a very rich repertory of subjects such as scenes of daily life, rituals and ascension to heaven. These examples of mural paintings contain a very complex interpretation code explaining the relations between life and death during the Western Han dynasty. A simple description of the subjects and also the 3D virtual reconstruction of the tombs are insufficient for approaching a correct cultural interpretation. In this paper we present a preliminary case study on the semantics of the tomb M 27's iconography (excavated in Xi'an in 2004 and documented by laser scanning) obtained through 3D virtual cybermaps. The use of virtual-cyber mind maps emphasizes the interpretation of the spatial, religious and symbolic connections (affordances) of the different subjects and images decorating the vault and the walls of the tomb. Through this simulation process the potential semantic recomposition of the tomb creates new metaphors of learning and communication.

***Key words**: 3D data collection, 3D reconstruction, cybermap, communication, learning process.*

1. Introduction

The Xi'an University of Technology Mural Tomb (M27) is an important monument discovered in the city of Xi'an and excavated in 2004. In June 2008 members of the University of California Merced and the Italian CNR (Bational Research Council) VHLabs (Virtual Heritage Lab) went to Xi'an and could survey the monument using a laser scanner. The mission, directed by Maurizio Forte, was possible thanks to an agreement between UC Merced and the Jiaotong University of Xi'an, aimed to the digital documentation of Western Han monuments for preservation and communication pourposes. The outcome of the project will be the creation of the *Western Han Dynasty (206 BC-9 AD) Virtual Museum* based on collaborative environments, and able to integrate new archaeological datasets coming from the fieldwork activities, monuments, and famous collections of artifacts of the Xi'an archaeological museums.

M 27 is one of the few Western Han Mural Tombs discovered in the city, which was the capital of the Western or Former Dynasty under the name of Chang'an. Despite its importance, the preservation of this heritage is at risk. In particular paintings inside the tomb are disappearing because of the critical conditions of plasters and colours. Since they are made over a very thin stratus of clay, directly applied on the bricks composing the structure, their removal from the walls and preservation inside a museum is impossible.

This is not the only case. Every year, in fact, chinese archaeologists in Xi'an discover hundreds of tombs. The monuments are mainly discovered during emergency surveys in construction sites. In the last few years the city has rapidly increased, and archaeologists are forced to collect quickly all the information on the ancient sites before them to be covered again, and disappear under a modern building. M 27 is one of the fortunate tombs to be physically preserved. But, unfortuately, the lack of economical resources prevent both a good preservation approach, and the public fruition of the monument. For all these reasons the VHLab's work is of extreme importance, because it permits the preservation of the monument through the digital documentation. This is not the only important aspect of the data-collection fieldwork. The data have already been processed in the lab, and the outcome has been the 3D virtual reconstruction of the tomb. If the model is useful for preservation aims, it is not enough when talking of musealization. Without a good infrastructure for showing the original context of any monuments, every 3D reconstruction becomes an empty box. The virtual musealization doesn't differ from the real. Every archaeological site, every museum, need infrastructures useful for the comprehension of ancient contents.

A common problem, when thinking to heritage communication, is the traslation of all the information in a comprehensible and explanatory musealization system, that could help visitors to interpret ancient codes. For the M27 case-study we tried to create this infrastructures working on the idea of 3D cybermaps. With the term cybermap we define the 3D graphic translation of those spatial connections.

Figure 1. The Xi'an University of Technology Mural Tomb (M27).

We started from the analysis of the paintings, that show a very rich repertory of subjects, such as scenes of daily life, rituals and ascension to heaven, containing a very complex interpretation code, which explains the relations between life and death during the Western Han Dynasty. We think that this contents can be very well understood stressing the spatial relations between every scene. Visiting the tomb we immediately perceived that the orientation and the location of every single scene was of extreme importance. We are convinced, in fact, that in this monument there is a corrispondence between spatial and conceptual relations.

The cybermap has been presented in a short movie in which we explored the idea of a learning process obtained through different layers of perception: the virtual recostruction of the tomb, the same model texturized with archaeological sketches, the cybermap. We are also working on meta-data; the final goal should be the creation of an interactive cybermap.

In conclusion our work has stemmed from two queries: can 3D data-collection increase the preservation capacities of our cultural heritage?

Can we use these data and different digital tools to improve the communication and the learning process in the museum, and how?

To answer to these questions we developed a project divided into several specific tasks:

- tomb 3D relief, realized using laser scanner tecniques;
- high resolution 3D model of the monument, useful for both preserervation and communication purposes;
- cybermap;- short movie to present in public, as the actual output of this project.

F.G., P.D.G., N.D.

2. Xi'an University of Technology Western Han Mural Tomb

Xi'an University of Technology Mural Tomb is located on Leyou Plane, northwest of Yue Jiazhai Village in the south suburb of Xi'an (fig. 1,).

Originally it had to be surmounted by an earth mound on the top, that was cut during the 1950s and 60s. The mural tomb was excavated between January and March 2004.[1] During the excavation robbery holes were found in the north wall of the tomb chamber, and on the brick wall closing the main chamber entrance.

The monument is characterized by the following elements: a tunnel excavated directly in the ground; two side rooms, where grave goods were preserved; the main chamber. The orientation is south-north as Chinese houses were traditionally set up in that time. Through the tunnel is possible to have access to the three underground chambers, that are all made of bricks.

The side rooms are located at the end of the tunnel, before the main chamber gate. In the eastern side room four seals and one pottery basin were discovered. In the western room two lacquer wooden chariots were placed in the front and back position. Both were rectangular in plane and had double shafts and single horse. Of these chariots only the lacquer pieces and the bronze ornaments remain, while the wooden part has deteriorated into wooden powder.

The main chamber has a rectangular shape (4.6 meters long from the south to north and 2.08 meters wide from the east to west). The chamber wall is 1.25 meters high, and the chamber itself is 2.10 meters high. It was completely covered of frescos. Inside the main chamber, the coffin of the deceased was found located in the north-east corner (2,10 meters long and 0,60 meters wide). Next to the coffin there were two jade eye masks, a jade object outlining in the mouth of the deceased, two jade nose elements, and a jade necklace ornament. Moreover 200 Wuzhu coins and a bronze seal were found in the chamber.

The tomb was dated at the end of the Western Han Dynasty (206 BC-9AD) thanks to a typological comparison with similar structures that were dated, and to the bronze Wuzhu coins discovered inside the tomb. The identity and status of the tomb owner have been assumed based on the size of the tomb, the importance of the burial objects, and the paintings. He was probably a governmental official during the Western Han Dynasty.

After the establishment of PRC (People's Republic of China), with the economical development of China,

[1] All the information on this tomb comes from the excavation report of the tomb: VV.AA. "Excavation Report on West Han Mural Tomb on the Campus of Xi'an University of Technology", *Cultural Heritage* [J], (Issue 5). Xi'an Municipal Institute of Cultural Relics Conservation & Archaeology, 2006. The book was translated in an English inedited version by Lizhi Zhang, Jaotong University of Xi'an.

building works have increased and archaeologists have found a big number of Han Dynasty Mural Tombs. Over sixty of them have been reported through media, but mostly they are dated back from the period of Xin Wang Mang (9 AD - 25 AD) to the end of Eastern Han Dynasty (25AD - 220 AD).

Few Western Han Dynasty Tombs have been excavated so far and, according to incomplete statistics, one is dated at the beginning of the dynasty and eight at the end. Only four are discovered in the region of Chang'an, ancient capital of Han Dynasty, partially corresponding to the modern Xi'an. Their typology and size are similar, but they differ for style, layout and contents of mural paintings. The paintings in the Xi'an University of Technology Mural Tomb are very innovative. They are richer from an ichonographic point of view, and more structured and elegant. The innovation consists in the introduction of scenes of daily life, which become very popular during the Eastern Han Dynasty. In this latter period, in fact, there is a shift from the absolute depiction of celestial figures to the expression of ordinary people, considered the switch from visionary romanticism to rational realism.[2] According to Chinese scholars, this tomb together with the XJTU Mural Tomb, has bridged the gap concerning the Western Han Mural Tombs in the central plain of Shaanxi Province, gaining significant academic values in offering valuable references for the study on Western Han mural paintings.

P.D.G., F.G., N.D.

3. Mural paintings description

The mural paintings inside the main chamber of the Xi'an University of Technology Western Han Mural Tomb, are realized on a thin white clay layer applied on the walls and ceiling. Because of the nature of the murals, there are important conservation issues that need to be addressed. Even though the mural tomb is closed to the public, and the temperature and humidity are being controlled, timely anti-molding and anti-germ processing have completed, the successful conservation of these paintings is still at risk. They cannot even be detached from the bricks and collocated into a museum, because of the way they have been realized. All the paintings are characterized by a black ink contour line, filled up with natural colors. The main colors are green, white, yellow, red and black, and they have a symbolic meaning linked to the universal elements of chinese philosophy.

Entering in the main chamber, the first subjects that appears in the eastern and western sides of the gate are the two tomb fairies guardians, the dragon and the tiger (fig. 2). They are both in the standing position between clouds, holding a long narrow flag; the dragon body traces an "S" and it is characterized by a sharp mouth, round eyes, double horns, and golden scales; the tiger is winged with black stripes on the back, and the face is not recognizable.

[2] He Xilin 2001.

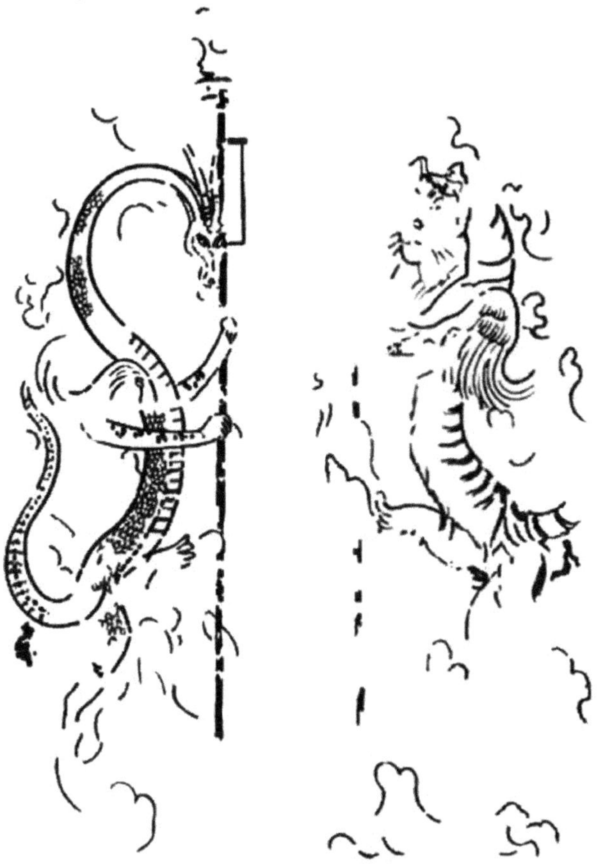

Figure 2. Side walls of the chamber gate (south): the tomb guardians. Archaeological drawing.

A second representation of a dragon and a tiger is on the ceiling, respectively in the south-east and in the north-west angle. The green dragon is connected with the sun in front of it -inside which there is a golden crow- and symbolically associated, in the chinese philosophy, to the East and the day. In the opposite angle the white tiger stands linked to the moon -in which there are a toad and a jade rabbit- representing the West and the night.

In association with these two fairies, in the middle of the south part of the ceiling, is the red bird of the South (zhuque), which seems to fly in the south direction. The zhuque, or red bird, is sometimes mistaken for the fenghuang, but they should be two separate entities. Fenghuang is a fantastic animal made up of the beak of a rooster, the face of a swallow, the forehead of a fowl, the neck of a snake, the breast of a goose, the back of a tortoise, the hindquorters of a stag and the tale of a fish.[3] It represents both male and famale entities as other symols in Chinese culture.

南方朱雀, Nán Fāng Zhū Què literally means 'the South red bird'. It appeared in association with the other three animals during China's Warring States period (476 BC

[3] Chang 1983: 56.

- 221 BC), and they were frequently painted on the walls of early chinese tombs.[4] In M27 the bird predominant colour is red, but it is not the only one; the body in fact is yellow, the peacock tail is multicoloured, the long neck is blue.

The last cosmic symbol is represented by a snake or a tortoise. They are strangely not depicted on the ceiling, but we can recognize two snakes on the northern wall, where there is the symbolic representation of ascension to heaven for immortal life.

On the ceiling it is also possible to recognize three celestial cranes among clouds. All the celestial birds and auspicious animals are flying to the south. In conclusion it can be said that the ceiling reveals the heaven in which the soul of the tomb owner lives.

On the side walls murals are featured by scenes of daily life, which underline the social status of the deceased:

- On the eastern wall scenes of horse hunting are represented. On the top of the south end of the wall there are chariot outgoing scenes: the master seems to be represented sitting in the two-horse leading chariot escorted by two riders with running horses opening the way, followed by one rider in the middle. On the middle and the northern part hunting scenes are represented; all the figures seems in movement to the north. We can recognize nine groups of characters: a red-clad hunter riding a white horse and holding a bow in his left hand and pulling an arrow in the right hand to shoot two fleeing deers in the front; two riders riding shoulder by shoulder; a hunter in red riding a black horse, and pulling his bow to shoot a deer; a red-faced hunter in grey riding a dark brown horse; a hunter in green off the horseback to pick up the prey; a hunter riding a white horse, holding the rein in the left hand, and keeping the lash, to whip the horse forward at fast speed; a hunter in red running after a wild boar; a white-clad ride hunter holding a spear on the back on a white horse; a hunter dressed in yellow, riding a red horse, and shooting the prey over his back; other figures are mutilated and unrecognizable. In the lower part, the most damaged, the scenes are difficult to be understood and described. The only recognizable is another scene of chariot outgoing.

- On the western wall, paintings in the north are striped off seriously. It is assumed that those paintings should be the scenes of musical and dancing performance. In the middle of the wall is the scene of rooster fighting. In the south of the wall is the banquet scene with dances; some characters are sitting on a wooden couch with a screen on the back; others are enjoing the show sitting on the floor; at the centre of the scene there are two dancers. The screen is a very important element during the Western Han period, but more generally in Chinese culture. It represents a division between the private part of rich houses and the representative space.

On the northern wall is a Yuren -the celestial being who leads the dead to ascend to heaven for immortal life (plate 3b). He faces a dragon and the west, and his arms are stretching forward in the steady position, as if to hold the dragon snout. He has red face, huge animal ears, round eyes, high nose, protruded lip, curly and fluttering hair, and wings on the shoulder. The dragon is red in belly and green in the back, the head of which has been severely mutilated due to the robbery hole. Below the dragon are a yellow and a green snakes. In addition, there is another animal, which has striking eyes but unidentifiable trunk. Among those images clouds are painted.

F.G., P.D.G., N.D.

4. Tomb historical context

In the iconography of the tomb M 27 we can recognize the symbolic expressions of the Western Han culture.

The Han period marks the beginning of an imperial bureaucratic state, where the social status rests on on a rudimentary examination system and the growth of a land based aristocracy.[5]

During this period there was the first operation of imperial unity, in the sense of the continuity of a single dynastic house. Government was now more effective that ever been; cultural life was richer, with more frequent contacts with non-chinese people, a more sophisticated view of literature, and the embellishment of many of China's arts and crafts. In this period the religion had formed an integral element in society and politics.

The religious belief and practices of Han China are based on three major principles: that of the Five Phases, which regulated the cycles of growth, change and decay; that of the complementary forces of Yin and Yang; and that of the single overwriting presence of *tao*.

Five Phases

The term *wu hsing* is variously rendered as the 'Five Phases' or 'Five Elements' or 'Five Agents'.[6] In a 100 B.C. source it is said that the so called 'Yellow Emperor', Huang ti, who examined the movements of the stars, worked out their cycles and initiated the concept of the five phases that comprehend universal activity. The powers of the five elements became associated with materials (wood, metal, fire, water, earth) and other sets of objects or qualities that could be numbered in five, such as the colors (red, black, white, green and yellow) or the directions (south, north, west, east and center). The number five assumed a strong symbolism: there were five sacred mountains, five senses of human perception, five musical notes ect. Wu hsing were associated also with different animal symbols (green dragon at east, red bird at south, white tiger at west, serpent and turtle at north, no symbol at the

[4] Loewe 1970: 72.

[5] Ch'ü 1982.

[6] Loewe 2005: 38.

center). These animals marked the four seasons and four cardinal directions: the green dragon of the East (spring), the red bird of the South (summer), the white tiger of the West (autumn), and the black snake or tortoise of the North (winter). The fifth direction was the centre linked with no animal symbols.[7]

The four symbols appeared frequently as a decorative motif on the backs of mirrors, on porcelains and of roof tiles. It is important to notice that from the middle of the first century before Christ the Five Phases affected iconography. The four animals sometimes may be accompanied by what may be a fifth symbol, a mound.

Yin-Yang

According to one theory which was evolved at about 300 B.C., the creation of the world and the continued processes of nature were to be attributed to the complementary powers of two major forces of Yin and Yang. The different impact of these two forces could be recognized in the everyday phenomena of the world. Yin was associated with female, dark and cold, Yang with male, light and heat; and the rhythmical procession of natural phenomena depended on which of the two forces happened to be in the ascendant. Yin and Yang were manifested in types of energy or qualities, or the material elements of fire, water, metal, wood and earth, whose creation they had contrived but which were themselves powerful enough to ordain the form of the material world.

Tao

Tao is known in writings as the Tao-te ching and the Chuang-tzu as the rule that underliens the universe. Tao is also defined as the 'way', "is majestic, and brings into question the value of human assumptions, judgments and aspirations".[8] Tao theories had to be well known during Han Dynasty. According to Loewe, in a daoist school of thought the movements of nature had to be seen as advance and decline of Yang, followed by that of Yin, in a total of five phases (tab. 1).

Yang	rising growth	wood	green dragon
	maturity	fire	red bird
Equilibrium	tranquillity	earth	no symbol/ mound
Yin	rising growth	metal	white tiger
	maturity	water	snake/ tortoise

Table 1. Schematic representation of daoist interpretation of the nature universal movement. Table elaborated from Loewe 1982: 38.

To those who thought in terms of Five Phases and Yin and Yang, Tao was the order of nature within which those rhythms operated. Dynastic governments identified themselves to a colour and to an element as representative of its power. Han sovereignty was explained as representing the dominant element of water-black, or later earth-yellow, and the appropriate colours were chosen for ceremonial use and display. The yellow color became increasingly popular, so much so some emperors started to be named "Yellow Emperor".[9]

Another important aspect of the Western Han ethical and philosophical system is the introduction of Confucianism as official imperial doctrine under the emperor Han Wudi (156-87 BC). The aim of Confucianism was the creation of a complex system of social and moral laws to end the Chinese spiritual decay. Confucianism focuses on the human experience and never on supernatural and metaphysic aspects. For this reason it was the imperial instrument for the organization of the chinese system of emperor and government. Moreover Confucius thought the importance of ritual and music in shaping moral sensibilities, and he advocated the benevolent sort of hierarchical social order. The idea was a society where everyone knew their position and acted in accordance with it: "Let the rulers be rulers, ministers be ministers, fathers fathers, sons sons".[10] He was convinced of filial piety as the base of a good society. According to Grant Hardy and Behenke Kinney, if Daoism is associated with individualism, freedom, non-conformity, nature, retirement, wit, and mysticism, Confucianism is associated with hirarchy, order, social responsbility, service, and conformity.[11]

At this period the Chinese society could be classified in emperor, officials and non-officials, organized in a pyramidal hierarchy (fig. 3). To be official meant to have one of the most important position in the hierarchy; officials were chosen originally in the base of hard qualifications, then more frequently thanks to sponsorship or favoritism.[12] Also if lineage was not a requirement for the official career, the proximity to the central government, or good family connections, played a high role in shaping the official advancement.

Officials were in charge of bureaucratic duties. They were divided in three main branches and this division dated back to the Quin Empire: civil, military, censorate.[13] The senior office of the civil branch was the chancellor, the most powerful man in the administration, working directly with the emperor to evaluate reports and set policy. Under the chancellor there were nine ministers, head of government departments as that of ceremony, the palace, trials, palace gards, transport, state visits, the imperial clan, agriculture and the lesser treasury.

[7] Loewe 1970: 118-121.
[8] Loewe 1982: 38-41; Loewe 2005: 43.

[9] Khon 2005: 136-137.
[10] *Analects*, 12:11.
[11] Hardy and Kinney 2005: 5-6.
[12] Lowe 2006: 71-85.
[13] Hardy and Kinney 2005: 31-33.

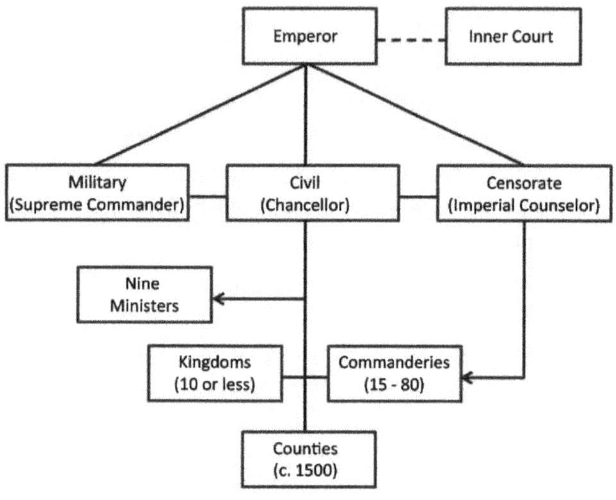

Figure 3. Western Han governmental structure. Elaboration from Hardy – Kinney 2005: 32.

At the beginning of Han Dyasty (Western Han) one third of the empire was organized into fifteen commanderies, while the rest of the land was divided among the kings. However the number of kingdoms decreased in favour of new commanderies that became about eighty in total. Each commandery was divided in counties and was headed by a civil governor and a military commandant.

During the Han Dynasty the cult of immortality was also important and it influenced religious practices. The world of the immortals is sometimes depicted in the fresco paintings of Han tombs. The Yuren, a celestial being partly human partly animal, was considered the spirit who leads the dead to ascend to heaven for immortal life. The desire to prolong life was not just linked to an individualistic conception of life, but to the desire of ensuring lineage and community continuity or survival. There were both the desire of a physical immortality and the conviction that was possible to obtain the bliss in a non-worldly immortality; the latter was proper of ascetics and hermits.[14]

Both the definition of this complex political and social system and the diffusion of Confucianism in the Western Han Dynasty could have brought about the introduction of scenes of daily life in the tombs. They could be read representing the social status of the tomb master.

F.G., P.D.G.

5. Fieldwork and post-processing: from the 3D data-collection to the 3D modelling

FIELDWORK AND 3D DATA-COLLECTION

The 3D data collection of M 27 was realized through laser scanning. The laser scanner used for the project was the Riegl LMS Z390i. It permits an acquisition with an accuracy of 6 mm for a range of 1-400 m.

This technology allows the digital acquisition of tridimensional objects, as point clouds. The resolution setted for the acquisition defines the density of the point cloud and the detail of the representation. Every point is described by a spatial position in x, y, z coordinates respect to the origin represented by the position of the scanner.

The model obtained is characterized by high geometric resolution, that is fundamental for the correct interpretation of the structure.

The data acquisition plan is the most important step during the fieldwork. It is fudamental to choose the level of detail in advance, while planning for the result of the final data output. The 3D data capturing during the fieldwork depends, in fact, on the final project purpose.

A project finalized to the fruition of the 3D models in a virtual reality environment, for instance, requires the optimization and pre-processing of the high resolution models. A high number of polygons could decrease the virtual reality engine performances, because they work on a real-time calculation and loading of the model polygons.

The M 27 3D relief is strictly connected to the *Western Han Dynasty Virtual Museum project.* The 3D data collection of this tomb was realized with a high level of detail (8 mm), even if the final goal of the project, the fruition in a virtual reality environment, doesn't necessarily need a such a detailed acquisition. This choice is strictly connected with the huge preservation problems linked to this heritage at risk.

According to these arguments our project his necessarily focused on two different research lines: if on one side we are working on communication of data in a collaborative virtual environment, on the other side we can't loose sight of the preservation value of an accurate 3D digital survey, and this in accordance with the so called CASPAR project (Cultural, Artistic and Scientific knowledge for Preservation, Access and Retrieval). This is a large-scale pilot scheme focused on the implementation of a pioneering framework to support the end-to-end preservation lifecycle for digital information based on existing and emerging standards.[15] In other words, the aim of this international project is the creation of a huge archive of digital data, collected for preservation purposes all over the world. It will provide information to scientists and permit the consultation and the scientific analysys of a library of different kinds of data in real time on the web.

Even if this project is currently in its embryonic phase, it is very important to start thinking about this new way to preserve information, decreasing the loss of world heritage.

[14] Loewe 1970: 114.

[15] http://www.unesco.org/science/remotesensing/?id_page=98&lang=en

POST-PROCESSING AND 3D MODELLING

The first step of the elaboration process of M 27 point clouds was the filtering. In order to obtain reliable 3D models the smoothing of point clouds is fundamental. It is necessary to smooth point cloud data in reverse engineering, or the inspection of free-form surfaces, because noisy points will have a negative influence on the post-processing of this data.[16] Different kinds of filtering were applyed (filter noise, filter redundancy, smooth points, sample points), without losing sight of the metric accuracy of the final 3D model.

The following step was the point clouds triangulation. This is a very delicate operation, whose success depends on the acquired data quality. In our case it was necessary to process two different point clouds on the basis of the project needs: the first one at high resolution was realized to archive the data for a preservation purpose. The second one was processed considering the data communication; in this case it was necessary to optimize the point cloud before triagulating it. This process allowed the creation of a 3D model with a controlled polygons number, an ideal solution to increase the virtual reality engine performances.

The digital model was texturized using different maps, in order to provide different levels of perception. The first texture was obtained by the high resolution pictures of the tomb. It is a georeferenced map coming from different photo captures of the tomb frescos. The images were applied on the surfaces of the objects using 3dsMax "unwrap modifier", which is a special tool of mapping, which permits to apply the image directly on the mesh. With this technique it is possible to link the plain of projection to the position of the digital camera.[17]

In order to provide a second level of interpretation and perception, a graphic reconstuction of the fresco was realized using the drawings done by the archaeologists, georeferencing them on the regular texture.

The final level is the overlapping of the regular texture with the drawings.

Different 3D model outputs were realized for the Xi'an University of Technology Western Han Mural Tomb. The first one was a 3D short movie in computer graphics, able to explain the structure, the frescos, and all related iconographies. Finally the tomb was imported in a virtual reality system using an open source engine (Open Source 3D Graphics Engine).

F.G., N.D.

6. The cybermap

STATE OF ART

The M27 mural paintings contain a very complex interpretation code, which shows the relations between life and death during the Western Han dynasty. Both the simple description of the subjects, and the 3D virtual reconstruction of the tomb, are insufficient for approaching a correct cultural interpretation. In this paper we present a preliminary study on the reconstruction of semantics of the tomb's iconography, obtained through a 3D virtual cybermap.

The cybermap (or hypertextual map in three dimensions) is the graphic layout of a set of relations between each scene and its context. Interacting with it, users are able to acquire as more information as possible on the tomb iconographic apparatus. A cybermap could be also defined as a mindmap, since it is the geometric translation of a mindscape.[18] Thanks to an abstract code it is possible to represent a symbolic environment, using different kinds of metadata.

The first cybermap was realized in 2001, as part of the *Scrovegni Chapel Virtual Reality Project*.[19] This chapel is located in the city of Padua (North Italy), and is one of the most important masterpieces of the European art. It was painted by Giotto between 1303 and 1305 and commissioned by Enrico Scrovegni. The paintings were restored in 2002, because of serious problems of preservation; but the accessibility to the monument is still limited in numbers of visitors (twenty persons) and in visit lenght (fifteen minutes). Therefore a Virtual Reality System was needed for improving on the musealization and the understanding of the monument. The paintings represent stories of the Virgin and Jesus Christ, and the Final Judgment, all organized in registers over the walls.

The map was designed to show the spatial relations between scenes and the thematisms which they were linked to: 1. themes concerning the general meanings and the paintings symbolic contents (thematic reading, exegetical reading, reading according to the lighting of the walls); 2. themes concerning the pictorial style (the light, the colour, the composition, the space).[20]

NARRATIVES AND MAPS

In the case of our tomb, a cybermap make-up was felt to be more challenging because of the complexity of contents. The idea of this work originates from the fieldwork. Visiting the monument we immediately understood that there was a big difference in the conception and use of space for the realization. As previously stated, the paintings are realized on a white clay stratus which hides the material support, giving the sense of an immaterial whole with intangible boundaries: contents and frescos spatial relations.

We can argue that an understanding of the tomb contents is possible just showing that intangible boundaries.

[16] Xue-Chang Zhang *et alii* 2006.

[17] Galeazzi *et alii* 2008: 133-136.

[18] "The mindscape is a virtual landscape perceived and interpreted by mental maps"; Forte 2002: 95-108.

[19] Forte *et alii* 2002: 43-52.

[20] Ibid.: 5-6.

Through a simulation process the potential semantic re-composition of the tomb creates new metaphors of learning and communication.

The iconographic complexity of the tomb frescos needs a cybernetic approach that could permit the decoding of the single representations and their pragmatic relations. Thanks to the cybermap it is possible to underline the strong symbolism that springs from the scenes.

The tomb is the logic and practice result of the revolutionary historical moment in which it was built -the end of the Western Han Dynasty-, and its paintings partially narrate and describe this period. They are visual narratives composed by scenes and themes.

It is well known that the human brain responds directly to the inability to process all the visual elements of a scene simoultaneously. As our eyes move from one point to another, they recreate a continous narrative that is perceived by the brain as seameless whole.[21] According to this statement the cybermap acquires an added value. It guides visitors in a virtual tour, showing the main iconographic themes paths; therefore it helps people to recreate narratives, moving from one scene to another in the right sequence. We should think of human consciousness as emerging from the complexity of such optical narratives through cause-and-effect-models, graphs and timelines. For a century, the dominant view of the brain has been that of a simple "reflexive" organ. In this view, human brains are little more than input-output machines. But it has been demonstrated that neurons don't simply wait for incoming data to be activeted. They are always active and choose whether and how to respond to stimuli.[22] John J. Ratey argues that the brain is: "a powerful prediction machine, continously making elaborate mental maps of the world that are reliable enough to enable us to predict what lies ahead, both in space and in time".[23]

If the material monument represents the tangible remain of Western Han heritage, the frescos spatial relations are traces of its intangible one. The map schematizes the themes and simplifies the information.

SCENES, THEMES AND PATHS

The spatial analysis of frescos in M27 permitted to underline their semantic value, but also to understand the monument pragmatics.

The scenes described before are connected to four main themes:

Daily life (fig. 3a): as already said, this theme is rarely present in late Western Han tomb. It symbolizes the social status of the owner, and testifies the introduction of Confucianism as official imperial doctrine. The scenes are also symbols of earth and mortality. They are painted in the two lateral walls of the tomb. The path starts from the eastern wall (representing daily activities) and continues to the opposite (representing nightly activities).

Ascension to heaven (fig. 3b): the core of this theme is the Northern wall (opposite to the entrance), where the yuren stands waiting for the deseased for the immortal life transfer. The yuren is the means of this transfer. The trip symbolically starts from the entrance where, on the two sides, are guardians animals (tiger and dragon) defending the sacred place from intruders. It continues with the daily life scenes, then with yuren on the northern wall, and ends on the ceiling where the heaven is represented.

Five Phases (fig. 4a): this path is designed on the ceiling, where the red bird, green dragon, black snake, white tiger, design a map themselves, being symbols of cardinal points. The path shape is circular, and it describes a continuous movement, defining a circular conception of time (eternal repetition).

Yin and Yang (fig. 4b): following this path the tomb can be devided in two triangles, where the verteces are cardinal points: the South-East/North-East/South-West triangle represents the yang (light, male, day); the South-West/ North-West/ south east triangle represents the yin (dark, female, night). This idea is reinforced by an observation: as already described, the green dragon of the east is painted in direct relation with the sun, and the white tiger with the moon. While the red bird and the snake are perfectly at north and south, the east and west symbols are designed at the two opposite corners South-East and North-West.

Why this anomaly? We don't have any certain answer to this question. We discussed a lot with Chinese archaeologists we are collaborating with, and the question is still open. We would just propose a suggestion: that the position at the corners of the two symbolic animals, the sun, and the moon, is not accidental. We tried to superimpose a symbol of Tao to an unwrapped image of the frescos and we saw that the location of that elements can suggest the background of this philosophical symbol, also if it appears later in time. We shouldn't forget that the creation of a symbol is just the final, graphic result of a cultural process.

In the 3D reconstruction we translated every scene in a simple object (a cube), created 3D connections between scenes, and assigned a colour to each theme.

CONTENTS COMMUNICATION AND LEARNING PROCESS

The cybermap has been presented in a short-movie as part of a more complex process of contents layout. In a movie of three minutes we decided to "musealize" the tomb, showing different layers of perception. In other words we started the movie with the simple 3D reconstruction of the tomb; then we showed the tomb model texturized using the paintings drawings made by archaeologists; in

[21] This idea was developed by Vik Muniz in the exibition "Rebus" at the New York MOMA museum. He arranged art and design objects from the MOMA's collection in a special order, and asked visitors to create a narrative from these artifacts.
http://www.moma.org/visit/calendar/exhibitions/304.
[22] Ratey 2002: 110-114.
[23] Ibid.: 112.

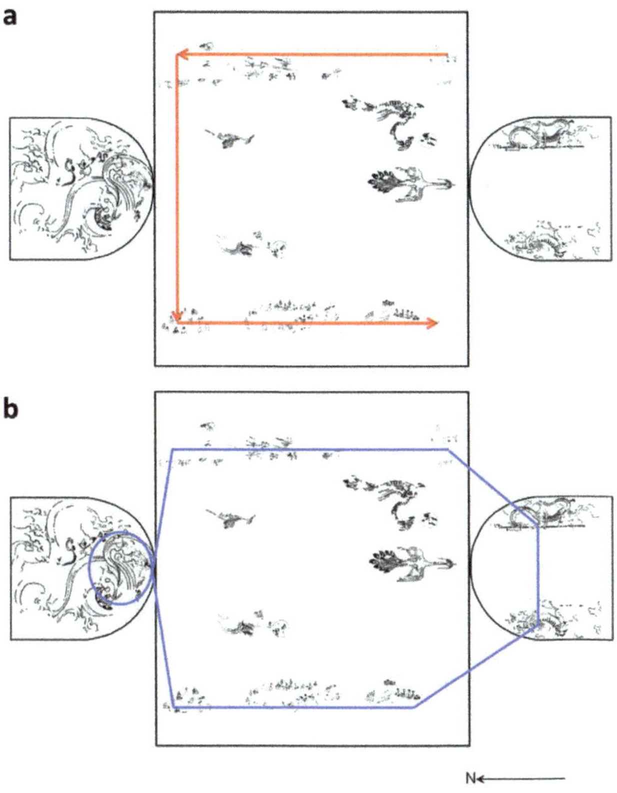

Figure 4. Themes: a. Daily life; b. Ascension to heaven.

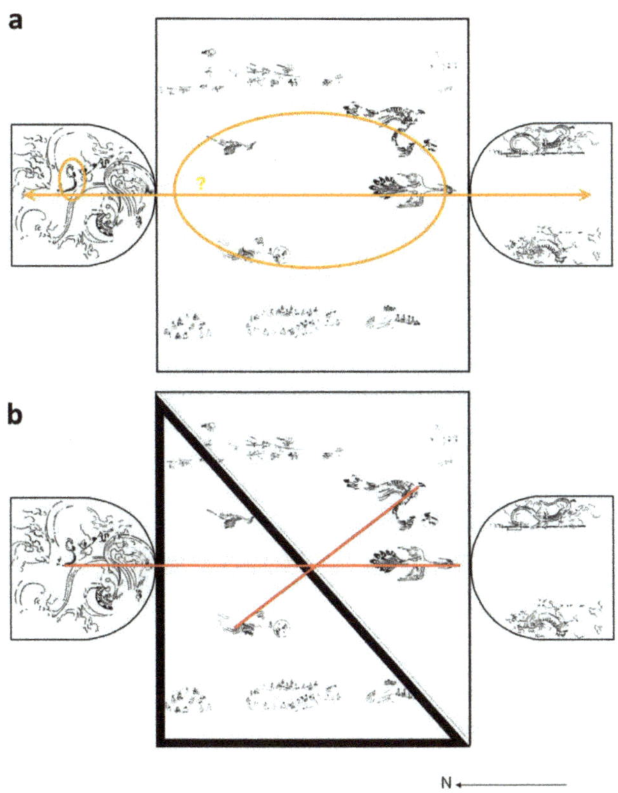

Figure 5. Themes: a. Five Phases; b. Yin and Yang.

a third step we overlapped the drawings to the texture of actual paintings, adding the cyber map; finally we reproposed the original 3D model (fig. 5). We conceived every layer as a cognitive model which helps the users to have added knowledge on the object. The viewer should be able to collect the information, step by step, and to retain it. The idea to end the movie with the simple 3D recostruction is based on the convintion that the user can recreate the fresco contents using information previously collected from the cybermap.

The human brain collects information thanks to mirror neurons. They are premotor neurons discovered in a sector of the posterior paretial cortex, that permit to map observed, implied, or heard goal-directed motor acts on their neural substrate in the observer's motor system; they also allow a direct form of action understanding, through a mechanism of embodied simulation.[24]

Through mirror neurons, the user is able to retain the map information and is able to mentally locate it in the original position in the 3D space. In this way it will be possible for the visitor of the virtual tomb to enjoy the paintings without any graphic (cybermap, drawings overlapped to the texture) obstacle, but with a new knowledge, previously acquired.

As already said, we are working on a complex system that includes meta-data. As a final result the cybermap will become interactive. It will be possible to link every cube to metadata, activating the learning process through affordances. According to Gibson the affordance is the quality of an object or an environment allowing individuals to perform an action[25]; "affordance consists in the opportunities for interaction that things in the environment possess, relative to the sensorimotor capacities of the animal".[26] They are stimuli for the brain to activate a learning process.

P.D.G., F.G.

7. Conclusions

In the Xi'anUniversity of Technology Mural Tomb case study a new methodology was tested, with a focus on the creation of new parameters in the archeological and communication field. The possibility to use new technological tools today, forces scholars to re-think and rearrange the old way to make archaeological documentation and musealization. Starting from the 3D data collection on the fieldwork, and concluding with the creation of a 3D cyber map, we tried to explain the added values of this new approach. We think that increasing the number of perception level of heritage through technologies, it is possible to improve the user capacity to collect information, retain it and transform it in knowledge,

[24] Gallese 2009: 521.
[25] Gibson 1979.
[26] Varela *et alii* 1991: 91.

Figure 6. The four levels of perception of the tomb.

after a subjective interpretation. As already said, the output of this work has been a short movie were the perception levels were combined with the cybermap. The learning process linked to this approach can be described as an hybrid learning in between symbolic reconstructive and sensory-motor. According to Antinucci[27] the symbolic-reconstructive way to learn is based on textuals that are conceived as symbols. The user reads, decodes textual symbols, and eventually recreates in mind a visual idea from the textual description. In the sensory-motor learning process, he/she learns through perception and action on the reality; users perceive an event with senses, acts on objects and changes the perception on the event after the action. This second process is augmentative: the action can be seen as a cause producing an effect, which is a new action. If the short movie urges people to utilize senses, the absence of interaction avoid a full sensory-motor learning. The latter is the core of an immersive experience. The final aim of the project will be the creation of an immersive collaborative environment, where the cybermap will be interactive and linked to metadata such as visuals, images, textuals and videos. The idea is close to that realized in the Virtual Museum of the Villa of Livia. This museum is exposed inside the National Museum of Rome, and it is the virtual reconstruction of Augustus wife's villa (first century A.D.). It has been conceived as a virtual collaborative environment, where visitors can interact with the system, but also share information between them. They can visit the virtual, choosing an avatar that represents them in the 3D space.

In conclusion this research stemmed from two queries: can 3D data-collection increase the preservation capacities of our cultural heritage?

Can digital data ameliorate the communication and the learning process in the museum, and how?

With this reaserch it has been proved that an high resolution data capture has a huge potential for preserving heritage at risk. However the 3D model alone is not enough to show the original context of an ancient monument. Every archeological site, every museum need an infastructure for the general public. At the same time a virtual environment needs digital tools for users in order to understand ancient codes and be able to reconstruct the context in every monument (fig. 6).

F.G., P.D.G., N.D.

Acknowledgements

Special thanks are due to Xi'an Jaotong University in the person of Li Jiajun, Gong Yenyin and Li Wen. Special thanks are also due to the Xi'an Municipal Institute of Cultural Relics Conservation and Archaeology in the person of Cheng Linquan, Guo Yongqi, Zhang Xiangyu, Shao Zhenyu, Yang Yonggang; and to the Xi'an ICOMOS in the person of

[27] Antinucci 2004a: 17; Antinucci 2004b.

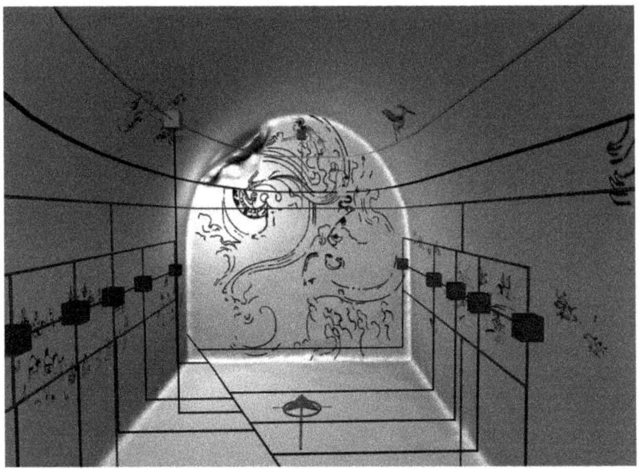

Figure 7. Hybrid representation of the tomb paintings.

Sun Fuxi. To professor Li Jiajun and doctor Cheng Linquan especially we owe acknoledgements, because of their scientific advise on our research, and for sharing their knowledge and all the material collected in the fieldwork.

We are thankful to Lizhi Zhang, in art Abby, for her efficiency, kindness, help in any kind of situation. Thanks to her we felt China as our country. We are also grateful to Mandy Chan, an incredible young and bright researcher, with the hope of future collaborations. Last, but not least, a special thanks to Emily Wilson for the review advises.

Bibliography

Analects, 12:11 Waley, Authur, trans. *The Analects of Confucius*, New York: Vintage Books, 1938.

Antinucci 2004a, Antinucci, Francesco. "Edutainment e beni culturali. Un introduzione critica". Valentino, Pietro A. – Delli Quadri, Maria Rita *Cultura di gioco*. Firenze: Giunti, 2004: 16-28.

Antinucci 2004b, Antinucci, Francesco, *Comunicare nel museo* (Percorsi), Milano, 2004.

Bateson, Gregory. *Mind and Nature. A Necessary Unit*. New York: Dutton, 1979.

Ch'ü T'ung-tsu. *Han Social Structure*. Seattle: University of Washington Press, 1972.

Chang K. *Art, Myth, and ritual*. 1983.

Forte, Maurizio. "Mindscape: ecological thinking, cyber-anthropology, and virtual archaeological landscapes". Forte, Maurizio – Williams, P.R., *The reconstruction of archaeological landscape through digital technologies*, proceedings of the 1st Italy-United States workshop (Boston, Massachussets, USA, November 1-3 2001), BAR International Series 1151 (2002): 95-108.

Forte, Maurizio – Pietroni, Eva – Rufa, Claudio. "Musealising the virtual: the Virtual Reality project of the Scrovegni Chapel of Padua". *Creative and Digital Culture*, proceedings of the International Society for Virtual Systems and Multimedia (Gyeonggju, Korea, 25-27 September 2002), Seoul: VSMM 2002: 43-52.

Galeazzi, Fabrizio, Moro, Alessia and Vico, Lola. "La modellazione architettonica. Il texturing". Forte M. (ed.) La *villa di Livia, un percorso di ricerca di archeologia virtuale*. Roma: L'Erma di Bretschneider 2008: 133-136.

Gallese, Vincenzo. "Mirror neurons, embodied simulation, and the neural basis of social identification". *Psychoanalytic Dialogues*, 19 (2009): 519-536.

Gibson, James J. *The ecological approach to visual perception*. Boston: Houghton Mifflin, 1979.

Hardy, Grant and Kinney, Anne Behnke. *The establishment of the Han Empire and Imperial China*. London: Greenwood Press, 2005.

He Xilin. *Ancient Tomb Paintings: Discovery and Study of Han Dynasty Tomb Chamber Murals[M]*. Shaanxi People's Press, First Edition, Nov. 2001.

Kohn, Livia, *Daoism Handbook*. Boston: Brill AcademicPublishers, 2005.

Loewe, Michael. *Everyday life in Early Imperial China*. New York: Dorset, 1970.

Lowe, Michael. *Chinese idea of life and death: Fate, myth, and reason in the Han period (202 BC-220 AD)*. London: George Allen & Unwin, 1982.

Loewe, Michael. *Faith, Myth and reason in Han China*. Indianapolis and Cambridge: Hackett Publishing Company, 2005.

Loewe, Michael. *The Government of the Quin and Han Empires (221 BCE-220 CE)*. Indianapolis and Cambridge, Hackett Publishing Company, 2006.

Ratey, John J. *A user guide to the brain. Perception, attention, and the four theatres of the brain*. New York: Vintage Books, 2002.

Varela, Francisco J. , Thomson, Evans and Rosch, Eleanor. *The embodied mind: Cognitive science and human experience*. Cambridge, MA: The MIT Press 1991.

VV.AA. "Excavation Report on West Han Mural Tomb on the Campus of Xi'an University of Technology", *Cultural Heritage* [J], (Issue 5). Xi'an Municipal Institute of Cultural Relics Conservation & Archaeology, 2006

Xue-Chang Zhang, Jun-Tong Xi and Jun Qi Yan. "A methodology for smoothing of point cloud data based on anisotropic heat conduction theory". *The International Journal of Advanced Manufacturing Technology*, Vol. 30. Numbers 1-2, London, 2006.

http://www.unesco.org/science/remotesensing/?id_page=98&lang=en

http://www.moma.org/visit/calendar/exhibitions/304

Cyber-archaeology and Virtual Collaborative Environments

Gregorij Kurillo,* Maurizio Forte,** Ruzena Bajcsy*
* University of California, Berkeley
** University of California, Merced

Abstract

We present a prototype collaborative application for cyber-archaeology which is aimed at remote real-time interaction with 3D archaeological models through a shared virtual environment. At each location, a set of stereo cameras captures 3D video of the user in real time to create his/her avatar. The 3D data is sent over the network to the remote location where it is integrated and rendered in the shared virtual environment. The teleimmersive aspect of this work provides a novel approach to interaction and interpretation of 3D archeological models by facilitating immersive experience in collaborative setting of remote users. The framework is aimed to facilitate in the future the study and analysis of a virtual reconstruction process in archaeology with the help from virtual community to re-contextualize and reassemble spatial archaeological data.

Keywords:

Cyber-archaeology, 3D modeling, shared virtual environments, collaborative work, tele-immersion

Introduction

Significant portion of the digital documentation process in archaeology is based on 2D maps, photos, layers, GPS points and several kinds of metadata (dbases, texts, movies, reports, etc.). The introduction of geographic information systems (GISs) has partially modified in some way the systematization of the process of knowledge and data-input. GIS technology captures, stores, analyzes, manages, and presents data that are linked to a specific location. These spatial data can be retrieved based on geographically based queries; however, majority of the data is in two dimensions, integrating maps, surfaces and different spatial layers. The use (still sporadic) of 3D technologies started with the introduction of 3D laser scanning techniques in combination with 3D modeling and photogrammetry (or, for example, computer vision based photomodelling) (Forte, 2006). This type of applications in archaeology has significantly increased in the last years, at different scales and with different case studies (Forte 2008). Monuments, tombs, cities, landscapes are typically recorded by time of flight scanners (accuracy under 1 cm); with the most powerful scanners in this category achieving hundreds thousands points per second at a distance of hundreds meters. Smaller monuments, various architectural elements and artifacts are recorded by optical and triangulation scanners (accuracy in a range of microns and a field of view of 50-200 cmq). Availability of such 3D data opens new possibilities to analyze the findings in the spatial format with respect to geometry, shape, volume, texture, coordinates and location.

Nowadays the 3D technologies of documentation and reconstruction are able to produce a relevant amount of data in a very short time and they can have a significant impact in the interpretation process of the archaeological research. The bottle-neck, however, is in the post-processing and the interactivity with the data. This is a very complex issue since the post-processing is a very time consuming activity and the use of data and models is highly limited with a scarce impact in the archaeological community. In fact there are very few significant examples of new interpretations in 3D domains.

In the last decade we have seen an increase in technological applications specifically in three domains: computer graphics (movies, games), visualization (VR, Caves, immersive systems, image processing), and virtual collaborative systems (interactive sharable systems). In computer graphics, novel (real-time) rendering algorithms combined with fast graphics processing units (GPUs) allow for creation of realistic virtual environments. 3D display technology has also made great advancements in quality and popularity in the recent years, shifting from displays intended for high-end virtual reality systems to commercial 3D television. Advancement in the networking technologies have facilitated creation of collaborative environments (mainly for gaming or social interaction) where users can interact with other geographically distribute users in real time. Such collaborative systems have had, however, limited exposure in scientific or academic communities. In particular the use of collaborative systems has been extremely limited in archaeology and within the archaeological community for technical reasons and, principally, for the scarce involvement of the scholars in the 3D interpretation process. In particular the simulation process is missing in this kind of research. It means that majority of the virtual reality applications are aimed at the reconstruction of virtual worlds and not to the spatial simulation and interaction of models, data, sites and artifacts. Use of 3D visualization and interpretation of an archaeological site from a macro (landscape) to a micro scale (stratigraphy) will create new perspectives of analysis and the experimental possibility to regenerate all the archaeological data in a virtual environment.

In our framework we aim to create accessible, sharable and validated archeological 3D content to assist with the interpretation process in real time using three-dimensional tools, spaces and interfaces: virtual worlds, experimental labs, and simulation environments for collaborative work. We plan to register and integrate different 3D and other data sources in the same space, similar to existing GIS systems, and provide a communication platform for real-time interaction of remote users. By combining 3D computer graphics, visualization and collaborative features we aim to create a virtual simulation environment where advanced behaviors, actions and new methodologies of research and training could be tested. Our framework also includes the teleimmersive technology which allows for realistic visualization of geographically distributed users in a shared virtual environment. Traditional immersive virtual reality systems often use avatars, to represent the human user inside the computer generated environments. In our work we move further from the avatars and apply stereo reconstruction to capture 3D representation of users in real time (Vasudevan *et al.*, 2010) to facilitate visual experience similar to reality (e.g. face-to-face meetings), where users are able to establish eye contact, recognize gestures, and communicate subtly via body language and facial expressions.

Related Work

As the virtual reality technology has become more widespread, there have been several attempts to recreate ancient worlds inside virtual environments. Initially, the applications were mainly aimed at visualization of data or the reconstruction results. For example, in the ARCHAVE project (Acevedo *et al.*, 2001) several research teams have explored immersive 3D visualization using Petra as a case study, but, because of display limitations, there was no room for collaborative work. In addition, these immersive systems were quite poor from the level of graphics, giving a disappointing sense of participation to the users and a few capacities of interpretation. Other projects, such as VITA: Visual Interaction Tool for Archaeology (Benko *et al.*, 2004), were totally oriented to the development of collaborative systems, integrating different types of 3D data. In the case of VITA two users wearing see-through, head-worn displays and tracked gloves were able to explore a portion of the virtual dig site using tangible interfaces and devices.

Lu *et al.* (2007) designed a computer-aided and collaborative system for the reconstruction of archaeological artifacts which allowed several archaeologists to remotely reassemble artifacts at the same time. Their research work mainly focused on algorithms assisting the reconstruction and support of multiple users rather than the immersive experience of the collaboration. Hall *et al.* (2001) explored a mixed-reality system SHAPE which allowed for exploration of archaeological artifacts through augmented reality in the public setting of the museum using head mounted displays. Several attempts (Harrison, 2009; Urban *et al.*, 2007) have also been made to use the existing collaborative virtual environments, such as Second Life, to feature results of the archaeological reconstruction processes or to display the artifacts in a virtual museum.

Recently, the use of 3D Web collaborative systems has been explored in the FIRB project (Forte and Pietroni, 2008), using Virtools DEV and Virtools Multiuser Pack© by linking three different archaeological sites. In this case all the collaborative activity was online and with pre-determined 3D graphic libraries. The advantages of this work are in the participatory interaction of multiple users sharing the same cyber space. Otherwise the limits of the system are principally in the low level of embodiment, the user interface (not very user friendly) and in the capacities of 3D interaction.

Although the massive multi-user environments seem appealing for such applications, the state-of-the-art of these environments currently does not provide users with truly immersive experience or enough flexibility to construct a complex framework that we propose. The users of these technologies were mainly observers of the virtual replicas of the ancient worlds rather than active participants contributing to the reconstruction and interpretation processes.

Most of the collaborative work in 3D virtual environments relies on computer generated avatars which cannot fully mimic human eye contact, gestures and subtle communication via body language and facial expressions. In our work, we therefore propose the use of 3D teleimmersion technology, such as the framework described by (Vasudevan *et al.*, 2010), to capture users in real time and integrate their 3D data into the shared virtual environment. The proposed framework will thus enhance the collaborative experience by providing more natural interaction between geographically distributed archeologists. In the past the presented framework has been successfully used in remote dancing applications (Nahrstedt *et al.*, 2007), learning of Tai Chi movements in virtual environment (Bailenson *et al.*, 2008), and collaborative work between geoscientists on volumetric 3D data.

System Overview

In this section we provide description of different components of the proposed 3D teleimmersive collaborative framework for cyber-archaeology (Figure 1).

Collaborative Virtual Reality Toolkit

Our collaborative application is build upon OpenGL-based Vrui VR Toolkit, developed by Kreylos (2008) at University of California, Davis. The Vrui VR Tookit aims to support fully scalable and portable applications that run on a wide range of virtual reality systems using different display technologies and various interaction devices. The toolkit provides abstraction between the physical devices and the virtual devices used by developed applications. The applications built with Vrui can run on various clients, from laptops to desktop servers, and support different

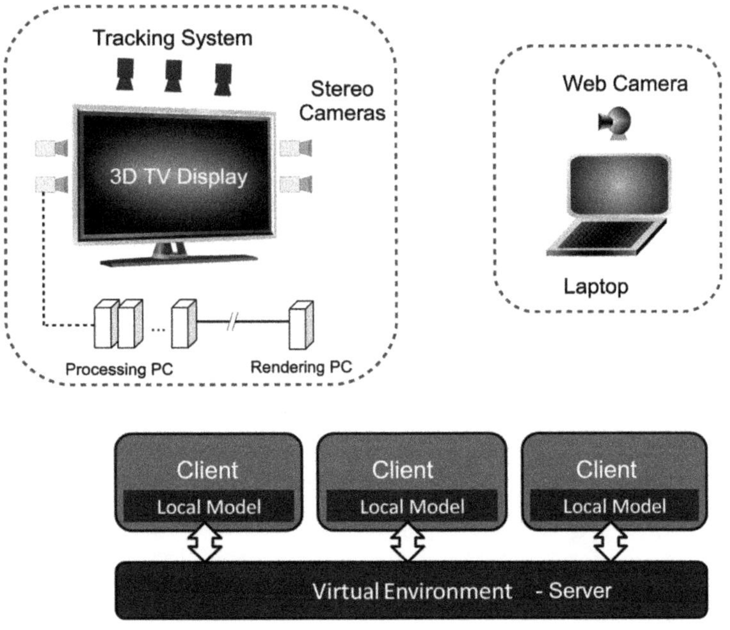

Figure 1. Components of the collaborative teleimmersion system. The framework supports 3D video capture with multiple stereo cameras. Data is processed by a computing cluster and the model is transmitted over the network and integrated in a shared virtual environment. The users with only a webcam are represented by a 2D video billboard to allow some level of visual feedback for communication with other users.

display technologies, such as 2D displays, stereo displays or fully immersive 3D displays (e.g. CAVE). The framework inherently supports several input devices and trackers with possibility to add custom devices without making changes to the developed application. The input device abstraction allows users to attach a virtual tool to each device and assign it with different functionality inside the application.

The collaborative extension of Vrui allows linking two or more spatially distributed virtual environments. Mirroring of input devices and cameras from one location to all the others is established through the collaboration datastream which is controlled by the framework itself. In addition, conversation data stream provides communication via audio, video or 3D video conferencing. Finally, application data stream updates application states between remote locations in real time (e.g. transmitting object location, locking objects etc.). The application data stream can be customized to control the behavior of connected applications.

Prototype Application

Using the above described tools we have built the first prototype of the collaborative application for cyber-archaeology which can be used on various display platforms. The framework currently supports use of various 3D models, shared virtual environment, real-time interaction, multi-media data streams for communication (i.e. audio, video and 3D video), and support for various input devices and display technologies (e.g. multi-view, stereo, 3D TV).

Shared virtual environment is implemented using a centralized scene graph which is handled by the server application (Figure 2). The scene graph allows for efficient and consistent rendering across remote clients. The collaboration and communication data streams have been derived from the Collaborative Vrui toolkit. Our custom application data stream consists of scene graph updates which are sent between the server and clients to allow efficient interaction with the virtual environment. The framework at this point is not intended to facilitate interaction of large number of users, such as for example in massive multiplayer games, but rather allow for interactive communication of small group of users (e.g. up to five).

Shared Scene Graph

Scene graph is used to describe the structure of the virtual scene and geometric relationship between different objects in the virtual environment. The scene graph consists of a collection of hierarchically organized, inter-connected nodes with parameterized spatial representation. Each node has one parent and it can have many or no children. The nodes are of several different types: (a) general node implementing the relationships within the scene graph (i.e. parent class incorporating node organization), (b) object node representing the geometry (i.e. vertices and texture coordinates), (c) transformation node defining the geometric relationship between connected nodes (i.e. transformation matrix), and (d) the root node.

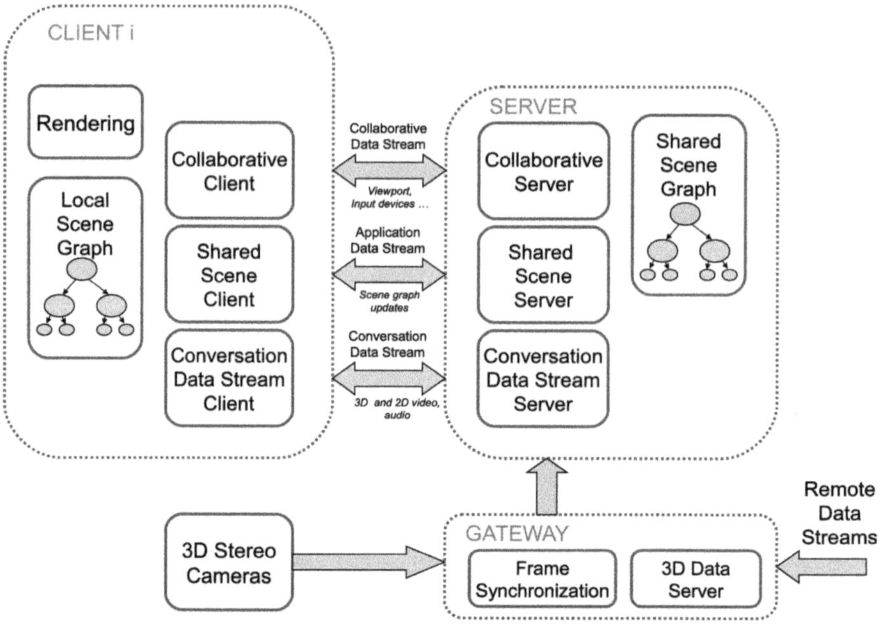

Figure 2. A simplified block diagram of the shared virtual environment for the collaborative cyber-archaeology with support for audio, video and 3D video communication.

In our framework we use two types of scene graphs, (1) shared (centralized) scene graph and (2) local scene graph. The shared scene graph is stored on the server side and describes the current consistent state of the shared virtual environment. The local scene graph is the last updated copy of the centralized scene graph and is used for efficient interaction and rendering of the object on the client side. This server-client-based model allows for synchronized interaction in the virtual environment. When a new client connects to the server, it receives a copy of the current scene graph which is stored as a local copy. The client loads the object geometry and texture into the memory based on the given scene description and renders it correspondingly as the geometry nodes. The clients currently support only OBJ/Wavefront 3D file format with possible extension to other geometry file formats and support for several texture file formats.

Changes made to the scene graph at the client side (e.g. moving an object) are transmitted to the server in real-time and broadcast to all the clients to update their local copies. The clients then render the updated scene while retaining consistency across multiple sites. Current prototype allows users to load, delete, scale, and move 3D objects in the virtual space or attach them to different parent nodes. These actions are in the form of short action messages transmitted from each client to the server, which in return broadcasts the updates (i.e. changes to the scene graph) to all the clients. With such an approach, the amount of transmitted data packages is minimal. The communication between the client and server is performed about every 20ms. If no changes are made to the scene, the application data stream contains only an empty message while no action message is sent to the clients from the server.

The scene graph is rendered locally using recursive calls from the root nodes to all the child nodes. Current implementation is using OpenGL lists which allow for interactive framerate (about 25 FPS) when displaying objects with about 1 million vertices. In the future the rendering module will be upgraded to GPU-optimized rendering (e.g. vertex buffer objects and vertex shaders) to handle objects with larger number of vertices.

3D Video Capture and Rendering

To capture real-time 3D video of users, we have integrated our system with the stereo algorithm presented by Vasudevan *et al.* (2010). The algorithm performs accurate and efficient stereo computation by employing fast stereo matching through an adaptive meshing scheme. In contrast to other stereo algorithms, this algorithm takes a hybrid approach: performing a local optimization technique (region matching) and using a global optimization approximation to improve the initial results (anisotropic diffusion). The algorithm therefore improves reconstruction of homogenous regions or regions with repeated patterns where other algorithms usually fail. It inherently produces a 3D mesh which is compressed and sent from each camera cluster to the local gateway. The achievable frame-rate is about 25 FPS on images with the resolution of 320x240 pixels. Several stereo views can be combined by externally calibrating the cameras to a common coordinate system to increase user workspace and to compensate for the occlusions (Kurillo *et al.*, 2009). A minimum setup for generating 3D video using this framework requires at least one stereo camera which can be mounted above the display. Depending on the camera properties and positioning, the camera may only reconstruct part of the user's body, for example the face and upper extremities, while still providing adequate feedback to enhance the communication channel between remote users. When several views are available, each view represents a partial 3D mesh which is streamed to the client renderer. The renderer based on the calibration data can fuse contributions of different views and generates appropriate image in the visual buffer.

Navigation

To create an interactive experience similar to traditional 3D applications (or games), each user navigates and interacts with the virtual environment in the first person perspective. At each location, the remote users are represented by their avatars rendered at their virtual location. As the remote user moves through the space, his/her avatar travels accordingly through the 3D scene as a part of the model space. If the user has 3D stereo cameras, the avatar is represented as their real-time stereo reconstruction (Figure 2). The users with only a webcam are represented by a 2D video billboard at their location to allow some level of visual feedback for communication with other users. The users who have no video acquisition system can still connect and interact in the shared environment as all the other users; however, their current virtual position will be indicated only by a 3D camera object. In the future, we plan to add ability to use customized 3D avatars in place. At any time, individual users can also switch to the collaborator's point of view or select face-to-face mode for direct conversation (Figure 3). The latter functionality will bring the local user in front of the remote user to facilitate a view similar to video conferencing interaction.

Interaction

As mentioned earlier, the Vrui VR toolkit provides the abstraction of the interface devices by mapping the device into virtual tools that can be assigned to device buttons beforehand or at run time. A wide selection of virtual tools is available to the users to interact with the application and objects:

- navigation tools: for navigation through 3D space
- graphic user interface tools: for interaction with menus and other on-screen objects
- measurement tools: for acquiring object geometry (e.g. dimensional and angular measurements)
- annotation and pointing tools: for marking and communicating interesting features to other remote users
- In addition to already available tools in Vrui, several custom tools were added to allow for direct interaction with the virtual objects:
- draggers: for picking up, moving and rotating objects
- screen locators: for rendering mode manipulation (e.g. mesh vs. texture)

Connected users can interact independently with the objects in the shared environment. To prevent inconsistencies, any two users cannot move the same object at the same time. A lock is placed on the node if a user is already interacting with the same object. The lock is assigned on first-come-first-serve basis. Since the local copy of the scene graph is always updated from the server, the user will not see any inconsistencies when trying to move already locked object.

To move an object, a dragging tool is used to first pick up the object and check for availability with respect to the node locking. Picking action is determined through simple collision detection algorithm between the dragging tool selection ray and pre-calculated object bounding box. If collision is detected, the object request is sent to the server and the user can start interacting with the object. If the object is already locked, the availability of the object is denied. The interaction with objects can be performed in all six degrees of freedom (DOF). When using a mouse to control the interaction, position and orientation in different directions can be controlled independently to compensate for reduced DOF (i.e. mouse is only a 2D device). More intuitive interaction can be achieved when using 6 DOF input devices. For such devices, tracking system has to be in use to acquire position and orientation of the device.

Once an object is moved from its initial position/ orientation, the new values are sent in real time to the server which consequently sends updates on the

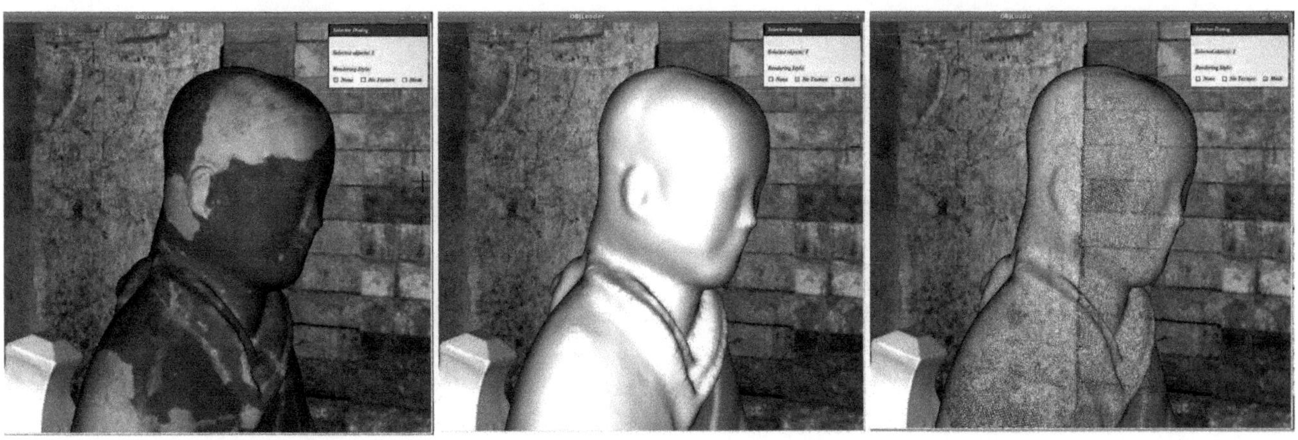

Figure 3. Different rendering styles are implemented to reveal the geometry of the imported models.

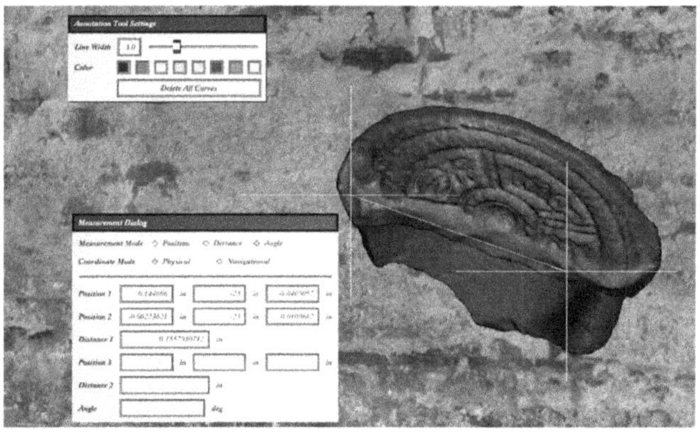

Figure 4. Dimensional and angular measurement can be performed on the objects to capture their geometry. Annotation tool can be used to mark important features.

corresponding node parameters to the clients. Remote users can in this way observe the real-time interaction with the object that the local user selected. With proper calibration (and hand tracking), the hand position of the remote avatar corresponds to the tool position in the virtual environment providing a realistic feedback on the interaction process. Figure 3 shows different rendering styles that can be achieved through the screen locator tool. Figure 4 shows the use of the measurement tool for capturing dimensions of objects and the annotation tool which is applied to mark important features in 3D space. Features annotations are transmitted in real-time also to the remote locations.

Hardware Overview

The proposed framework for the teleimmersive 3D collaborative cyber-archaeology can be used on various platforms. As mentioned earlier in the paper, the Vrui VR Toolkit, which the application is built upon, allows for scalability in terms of display and input devices. The minimum hardware consist of a laptop with a graphics accelerator, mouse input, microphone and speakers, webcam and wired or wireless connection to establish a 2D video and audio stream from the user into the virtual environment. Such a setup is appropriate also for field-work where other technologies are not available. On the other end, the framework supports more immersive environments which can include multiple stereo cameras to generate real-time 3D video of the user and various multi-displays or stereo displays. Users can easily switch between different hardware setups by loading different configuration file of their Vrui VR Toolkit installation.

In this paper we present results obtained using two different setups connected over the internet (Figure 5). For the first setup we used the teleimmersion platform at University of California, Berkeley (Vasudevan *et al.*, 2010) which consists of several stereo clusters, each connected to a four core server, to perform 360-degree stereo reconstruction. The system is integrated with a tracking system (TrackIR by NaturalPoint) which is used to track position and orientation of a Wii Remote (Nintendo). The Wii Remote served as a 6 DOF input device for interaction with the virtual environment. The second setup consisted of a single Bumblebee 2 stereo camera (Point Grey, Inc.) positioned above 65" LCD screen. Interaction was facilitated by a 3D mouse.

The framework is compatible with various stereo display technologies, such a 3D TV plasma display, which can be used in connection with active shuttered glasses and head tracking. The head tracking allows for realistic and immersive viewpoint-dependent rendering of 3D content.

Results/Outcomes

In this section new present some of the results obtained on the prototype system. More sophisticated interfaces, interactions and behaviors are planned to be implemented as described in the next section.

Nevertheless, at this stage, we did some preliminary experiments with the 3D archaeological data coming from a monumental Chinese Western Han Tomb (beginning of the first millennium AD) we recorded and documented by time of flight scanners in the summer 2008 at Xi'an, China.

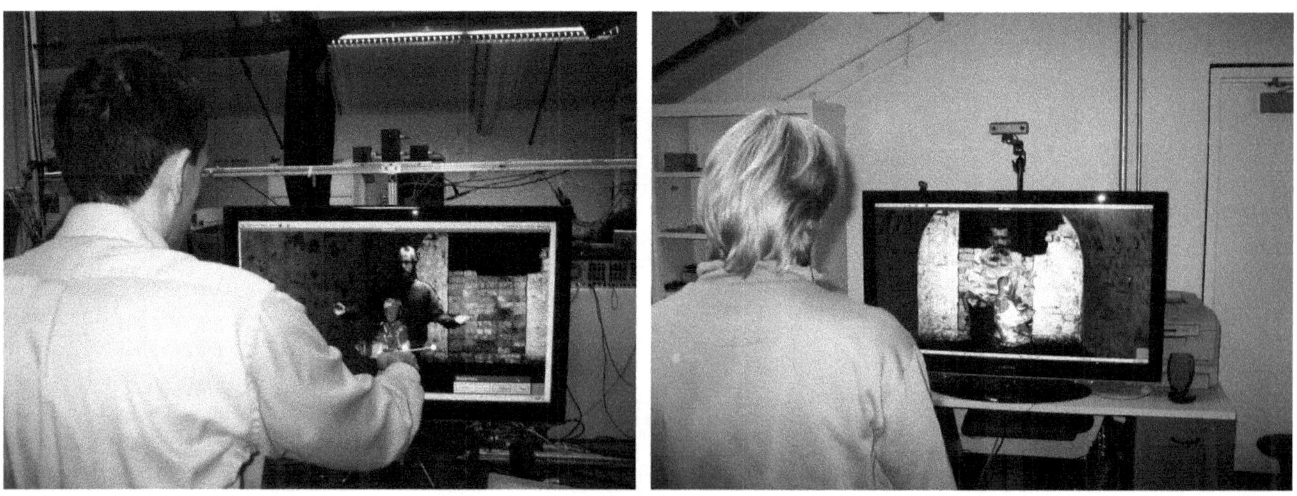

Figure 5. Two different teleimmersive platforms for interaction with the virtual environment. Left is the system featuring several stereo clusters and 6 DOF input device. Right is a simplified version with only one stereo camera.

The tomb, after the excavation, was closed to the public and is now accessible only virtually. In addition, most of the mural paintings were seriously damaged and they are virtually restored thanks to high resolution pictures obtained by the Chinese archaeological team at the time of the discovery.

In this case we have reconstructed in 3D the tomb and its funeral goods (recorded by laser scanning) in order to re-contextualize the artifacts in their original positions and analyzing them in the cyber-space (Figure 6). Even the corridor and the three chambers of the tomb have been studies within the collaborative systems; in particular the architectural elements, the organization of the space and the relation between iconography, the funeral chambers and the 3D model. The use of lighting and measurement tools in combination with the capacity to move, share and compare the objects in the cyber space, to add visual layers and outlines in the wall paintings have considerably increased the simulation factors and the faculties to interpret the data. The involvement of different interactors in the cyberspace disclosed new perspectives in the dialectics of the interpretation process and its multivocality. This kind of involvement/embodiment constitutes a kind of enaction where the interactions are more natural than in a game like experience and oriented to a deep real time comprehension and awareness of the surrounding environment.

The final framework will represent an open laboratory: a place where it is possible to compare the construction and validation of interpretative processes, to investigate new relations among data in space and time inside a shared virtual environment.

Future Developments

The objectives for this learning and simulation platform are to facilitate interpretation, reconstruction and communication of archaeological datasets through a remote virtual participation. We plan to realize these goals with two different, but interconnected direction of this framework, through building Spatial Database and use of Virtual Communities.

Spatial Database

The spatial database will integrate different data sources (i.e. 3D models and other spatial data, photos, movies, maps, and texts) into the same virtual space, similar to existing GIS systems. The representation will, however, be accessible in a 3D world with remote immersive interaction and communication. The virtual environment itself offers many advantages over the existing software which has been mainly used for visualization in the past but not the real-time interaction. Different laser scanned models will be spatial registered with respect to the 3D maps of the location (e.g. within the archaeological site). This will allow others to contribute to the interpretation of the findings by experiencing the site in an immersive virtual environment through 3D visualization similarly to

Figure 6. Locally, users observe the virtual world in the first person perspective. Remote users are represented by their 3D avatars captured by one or more stereo cameras.

a real-life visit of the site. The excavation could be also explored in the temporal domain (depending on the availability of such data), to display how different layers of the findings were revealed. The restoration process could be represented in a similar fashion. One can imagine layers of textures of a reconstructed fresco that can be interactively added, mixed or removed to reveal details that may not be apparent by comparing separate images on a 2D display. Interactive virtual viewpoint selection can also add value to the interpretation process by allowing archaeologists to explore the findings from different angles. We plan to also add the ability to control the lighting in the virtual environment. With complete 3D information available (i.e. vertices, normals and material properties), an artifact could be relit to reveal geometrical details, otherwise hidden from the observer. In addition to the 3D model geometry, the shared virtual environment in the connection with the spatial database will allow for display of other types of data spatially registered with the models (e.g. 2D photographs or maps could be displayed aligned with the 3D model for comparison).

The collaborative aspect of this work will provide experts in the field to work together remotely in the same cyberspace, interacting in real time with models of artifacts, monuments and sites and at the same time have access to other datasets within the environment. The users will also be able to bring up a portal to other applications running on remote computers through a VNC (Virtual Network Computing) plug-in. Users could for example connect to the web browser without leaving the virtual environment.

This prototype collaborative application for cyber-archaeology is aimed to demonstrate real-time collaborative interaction with 3D archeological models in connection with video streaming technologies, including light-weight 3D teleimmersion which provides a higher level of embodiment. The activity of learning will involve a bottom-up approach, the analyses of the archaeological remains and finds and a top-down approach, the reconstruction (for example architectural features, artifacts, frescos, styles, materials, shapes, and others). A VR collaborative domain is a simulation environment for testing advanced behaviors, actions and new methodologies of research and training. It could be conceived as an open laboratory; a place where it is possible to compare the construction and validation of interpretative processes, investigate new relations among data in space and time, and establish affordances by interactive ecosystems.

The study and analysis of a virtual reconstruction process in archaeology will help the virtual community to re-contextualize and reassemble spatial archaeological data sets, from the first draft version (data not yet interpreted) to the final communicative level which will be disseminated through the Virtual Communities.

Virtual Communities

In the final stage we are planning to connect our infrastructure through the Virtual Communities, such as OpenSim (opensimulator.org), and make the results available to a wider community of educators, students and others through the Internet. The integration will allow for creation of new processes of learning, research and education in virtual archaeological worlds, playing and interacting inside a 3D cyber-space. Such an environment will access the spatial database created through the collaborative virtual environment; however, due to inherent limitations of the 3D engine of OpenSim-like environments, the level of the interaction and immersion will be limited.

Different educational activities will be planned with regard to the virtual communities. Availability of the data to much broader audience will allow for transparent presentation of the reconstruction processes and interpretations.

In connection with our teleimmersion research, we are planning to create realistic dynamic avatars that could be integrated into these virtual communities alongside the existing avatars. This would provide users with stereo cameras to reach a higher level of embodiment while still allowing users with limited technological capabilities to interact in the same environment. Dynamic avatars would allow for users to express their emotions and body language more intuitively instead of using predefined scripts, for example to point to different features or initiate eye contact with other users.

Conclusions

Teleimmesive archaeology will allow researchers, faculty members, and students to work collaboratively on a variety of projects relating to cyber-archaeology, but also history, anthropology, social sciences. Cognitive scientists will conduct studies on how users will interact naturally with virtual characters in the virtual environment. Research questions in this vein will include the following questions: How do people interact with virtual characters and virtual humans, and how does this affect learning in a virtual environment? Is attention sustained and memory more robust for information about virtual historic objects (e.g., function of object, location of object) when virtual characters point at objects while they describe the objects? How will users as avatars grasp and manipulate virtual objects, and what are the cognitive benefits of this type of interaction? Two crucial aspects regard the role of "awareness" and "imitation". This provides the virtual communities with views or representations of each other and their work, to help them coordinate their actions smoothly. Awareness in collaborative systems may arise directly through the visibility of other people's action, or indirectly through the visibility of the effects of actions on the objects of work. The imitation factor concerns the capacity to create mental maps of someone's actions. The combination of "awareness" and "imitation" generates as outcome the social learning process, which constitutes the basis of any information unit and cultural transmission. The final virtual reconstruction process will make transparent any potential collaborative-participatory interpretation, creating new ways of research, training and communication in archaeology.

Acknowledgements

We thank Rama Vasudevan and Edgar Lobaton for the stereo reconstruction work at University of California, Berkeley. We also thank Tony Bernardin and Oliver Kreylos from University of California, Davis for the implementation of the 3D video rendering. Furthermore we acknowledge financial support from NSF grants 0703787 and 0724681, HP Labs, The European Aeronautic Defence and Space Company (EADS) and Center for Information Technology Research in the Interest of Society (CITRIS) at University of California, Berkeley.

References

Acevedo, D., Vote, E., Laidlaw, D. H., and Joukowsky, M.S. 2001. Archaeological data visualization in VR: Analysis of lamp finds at the great temple of Petra, a case study. *Proceedings of IEEE Visualization Conference*, 493-497. San Diego, California.

Bailenson, J. N., Patel, K., Nielsen, A., Bajcsy, A., Jung, S., Kurillo, G. 2008. The effect of interactivity on learning physical actions in virtual reality. *Media Psychology* 11, 354-376.

Benko, H., Ishak, E. W. and Feiner, S. 2004. Collaborative mixed reality visualization of an archaeological excavation. *The International Symposium on Mixed and Augmented Reality (ISMAR 2004)*, 132-140. Washington DC.

Forte, M. 2006. A digital "cyber" protocol for the reconstruction of the archaeological landscape: virtual reality and mindscapes. *Proceedings of the International Workshop on Recording, Modeling and Visualization of Cultural Heritage*, 336-351. London. Taylor & Francis Group.

Forte, M. 2008. Cyber-archaeology: an eco-approach to the virtual reconstruction of the past. *Proceedings of International Symposium on "Information and Communication Technologies in Cultural Heritage"*, 91-106. Ioannina, Greece.

Forte, M. and Pietroni, E. 2008. Virtual reality web collaborative environments in archaeology. *Proceedings of the 14th International Conference on Virtual Systems and Multimedia (VSMM 2008)*, 74-78. Cyprus.

Hall, T., Ciolfi, L., Bannon, L. J., Fraser, M., Benford, S., Bowers, J., Greenhalgh, C., Hellström, S. O., Izadi, S., Schnädelbach, H. and Flintham, M. 2001. The visitor as virtual archaeologist: explorations in mixed reality technology to enhance educational and social interaction in the museum. *Proceedings of Virtual Reality, Archaeology, and Cultural Heritage (VAST 2001)*, 91-96. New York.

Harrison, R. 2009. Excavating second life: cyber-archaeologies, heritage and virtual communities. *Journal of Material Culture* 14, 75-106.

Kreylos, O. 2008. Environment-independent VR development, in G. Bebis, et al. (eds.), *Advances in Visual Computing (ISVC 2008)*, 901-912. Berlin, Springer.

Kurillo, G., Li, Z. and Bajcsy, R. 2009. Framework for hierarchical calibration of multi-camera systems for teleimmersion. *Proceedings of IMMERSCOM '09*. Berkeley, CA.

Lu, Y., Gardner, H., Jing H., et al. 2007. Interactive reconstruction of archaeological fragments in a collaborative environment. *Proceedings of the International Conference on Digital Image Computing Techniques and Applications (DICTA)*, 23-29. Adelaide, Australia.

Nahrstedt, K., Bajcsy, R., Wymore, L., Kurillo, G., Mezur, K., Sheppard, R., Yang, Z. and Wu, W. 2007. Symbiosis of teleimmersive environments with creative choreography. Proceedings of ACM Workshop on Supporting Creative Acts Beyond Dissemination. Washington D.C.

Urban, R., Marty, P., Twidale, M. A. 2007. Second life for your museum: 3D multi-user virtual environments and museums, in J. Trant and D. Bearman (eds.), *Museums and the Web: Proceedings*. Toronto. Archives & Museum Informatics.

Vasudevan, R., Lobaton, E., Kurillo, G., Bajcsy, R., Bernardin, T., Hamann, B. and Nahrstedt, K. 2010. A methodology for remote virtual interaction in teleimmersive environments. *Proceedings of 1st ACM Multimedia Systems Conference*, 281-292. Scottsdale, AZ.

Comparing tangible and virtual exploration of archaeological objects

David Kirsh
UCSD

Abstract

Can virtual engagement enable the sort of interactive coupling with objects enjoyed by archaeologists who are physically present at a site? To explore this question I consider three points: 1) Tangible interaction: What role does encounter by muscle and sinew play in experiencing and understanding objects? 2) Thinking with things. What sorts of interactions are involved when we manipulate things to facilitate thought? 3) Projection and imagination. Archaeological inquiry involves processes beyond perception. Material engagement of things stimulates these processes. What must be present in a virtual environment to recreate the feel of material engagement? I conclude that nothing, in principle, prevents future virtual environments from supporting this material engagement of digital versions of artifacts. But, there is much that remains to be understood about how to realize this material engagement, both at a technological and a cognitive level.

Archaeological exploration is an embodied, material activity. To understand the function of physical fragments, archaeologists heft objects and feel their balance; they closely scrutinize them, they hold them at a distance, or place them against a background of other objects. Whether or not mental projection and imagination are deemed valid archaeological methods, they are still vital elements in inquisitive minds, and the physical act of working and playing with material objects stimulates both. Can we create enough realism with the digital objects populating virtual worlds to duplicate the cognitive feel of real artifacts? Can we re-establish the 'enactive' context needed to reproduce the cognitive processes that archaeologists enjoy in the field?

In focusing on the challenge that virtual archeology faces in recreating the cognitive depth of material engagement, I do not in any respect mean to question the special qualities that virtual archeology offers to archaeologists and the public. As others have discussed, virtual worlds allow us to communicate ideas at a distance, to collaboratively role-play with artifacts, to add or swap virtual backgrounds to contextualize our encounter, or to recreate elements of a dig forever lost. Because it is easy to augment any virtual scene with digital objects, it is easy to add computational elements that reduce the cost of information search, information visualization, and visual comparison. Scenarios can be created and history brought to life. There is little doubt in my mind that archeology will move increasingly in a digital direction both for archaeologists on site, who may wish to augment local reality, and for an enthusiastic public interested in virtual tourism and cultural inquiry.

FFrom a cognitive perspective, however, the question of how physical involvement with objects affects cognition remains an open and very active question. Many of us believe that by interacting tangibly with objects we get something more than by observing images of those objects, even 3D or holographic images. By interacting through a broad range of modalities – touch and manual manipulation in addition to sight and sound – we interactively couple with objects in ways that actually extends cognition. If it is true, then, as many have argued, [Clark 2003, Hutchins 2005, Kirsh 1995, 2009, 2010, Latour 1986)] that humans can literally 'think' with objects, how can virtual engagement support the sort of thinking with things enjoyed by archaeologists who are physically present at a site?

This question can be stated more bluntly. The practices of archaeologists in the field involve more modalities and more material engagement than are likely to be supported in virtual environments for some time. My objective here is to understand what more being in a physical world gives archaeologists.

To explore this issue I consider three points:

1. **Tangible interaction:** What role does encounter by muscle and sinew play in experiencing and understanding objects?
2. **Thinking with things**. What sorts of interactions are involved when we manipulate things to facilitate thought?
3. **Projection and imagination.** Archaeological inquiry involves processes beyond perception. Material engagement of things stimulates these processes. What must be present in a virtual environment to recreate the feel of material engagement?

Figure 1. Cutting tools from the New Stone Age.

Tangible Interaction

Look at the stones in figure 1. To an expert eye, the marks of artifice are apparent. Natural processes could, by chance, create such a thing; but the planes and facets have a rational structure that suggests purpose. How can one tell?

Two techniques help: hefting and drawing.

Hefting is the action of weighing and feeling the momentum of an object by moving it in a jerky manner. A stone made for chopping hefts differently than one made for cutting. Because of the distribution of mass moments, certain motions that feel natural with a chopping stone feel less natural with a cutting stone. This feel is only occasionally detectable by sight. Only sometimes can you tell by looking, whether an artifact will fit in the hand comfortably, and even less often can you tell which movements will *feel* most natural. Feel is sensitive to density and mass, and both are invisible. Sometimes affordances that are invisible to the eye are apparent to the working hand.

Today's virtual environments typically fall short on modalities such as touch, heft, smell, and so on. Pressure sensitive gloves, near-field haptics and more, are important input output devices in the laboratory. But hefting goes beyond standard virtual simulations of touch and feel. Hefting lets us perceive the distribution of weight and inertia. To recreate the 'feel' of an axe head, its mass at different points must be known.[1] Of course, this can in principle be achieved in virtual environments. But at present few, if any such environments do. They should.

The consequences of depriving a scientist of these observations goes beyond the phase of inquiry where a conjecture is being tested. It affects the conjectures

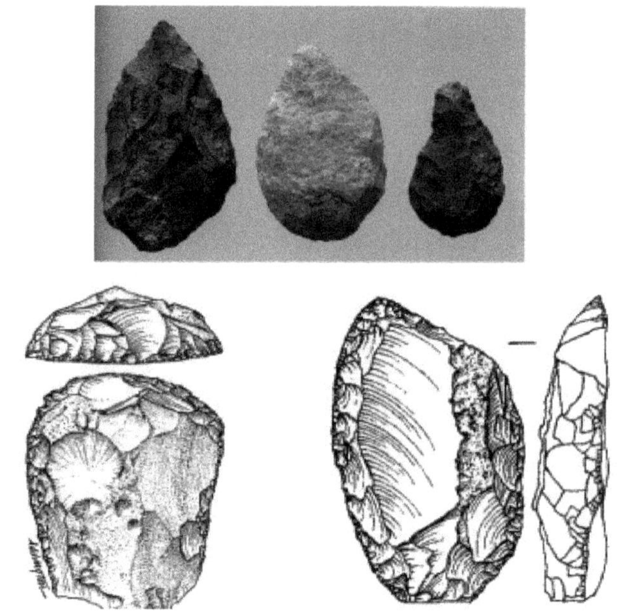

Figure 2. Three Paleolithic stones are shown in the upper part of this figure. Below are two examples of lithic illustrations, though of different stones. Note how much easier it is to identify the way the stones were knapped.

that are likely to come to mind. In a world without touch, we would expect fewer touch related hypotheses; fewer conjectures about the function of an object that are based on invisible modalities. In short, reduced imagination. Imagination is a topic we will return to later.

Drawing. A second, and rather surprising, method for determining whether a given Paleolithic stone is a cutting tool is to *sketch* the stone. See figure 2. Not just any form of sketching will do. There is an expert mode of sketching for Paleolithic objects codified in a set of principles of 'lithic illustration'.[2] Good archaeological illustrators will draw a lithic stone to reveal the physical 'problematic' the tool cutter faced. They will show the "scale; the pattern,

[1] In studies of motor perception, for instance, Carrello & Turvey (1996) found that subjects were able to determine the shape of a the information through hefting and wiggling a stick with a weight on it was sufficient to allow computing the inertia tensor of the object. That is, we are sensitive to the inertia of the mass at all points in the object (its mass distribution).

[2] Addington (1986).

sequence, direction, and force of blows to the stone; the bulb and platform of percussion; areas of retouch, snapping, and truncation; areas of grinding, battering, or abrasion; fractures caused by heating; the effects of materials; and pitting and sickle sheen."[3] Potentially confusing features of the stone such as embedded fossils, variegated coloration, patina, seams, banding, and crystallization are left out of the drawing.

The implication is that expert illustrators, when practicing their craft, are forced to scrutinize stones in a special way. They coordinate hand and eye to interactively probe the stone to reveal knapping related features. The need to draw certain lines drives perceptual inquiry. Attention must be managed, and arguably, without the need to sketch, without the presence of an external structure that the illustrator is creating, attention would not be managed adequately. Of course, this is an exaggeration. Illustrators have professional vision[4] and so can see elements of what they would draw without actually drawing. But in drawing, the process of making lines and ensuring they are spaced revealingly, is itself a process that simulates knapping. Using a pencil to draw a curve is physically related to using a knapping stone to flake a chip off a stone. It physically simulates knapping. So, the drawing process can help the illustrator walk through the history of the axehead's making. The drawing is an external representation, and the process of making this representation is a powerful method for structuring attention. It helps the illustrator to figure out what an artifact is by studying 'the details of its making' (ibid).

Simulating behavior. Illustration is a method of simulation that works indirectly by managing attention. A more direct method of understanding an object is to work with it manually. A third reason why tangible interaction adds something to perception *simplicitur* is that without physical interaction it is much harder to tell the multi-sensory ways an artifact might be engaged in a culture's practices. The role objects play in a culture is constantly surprising.

For instance, in figure 3, a person is shown rubbing a Chinese water bowl. When the metal handles of the bowl are rubbed the vibrations cause standing waves in the water. The creators of the bowl engraved four dragons on the base plate to make it seem as if they are spitting water. Now imagine you discover such a bowl at a dig. You see the bowl as a portable container that would probably hold liquids without leaking. But why the dragons, and why the large handles? They seem larger than is necessary to facilitate carrying. Might they play a role in a ritual?

How would one decide? The usual method calls on knowledge about cultural practices at the time, about religion and social rites. Naturally, it is vital to study the details of the room in which it was found, what was its

Figure 3. Water springs up from the mouths of four dragons, engraved on the bottom of the bowl.

location in the room, what was nearby it. All these things help to support one view or another. But why would one ever suppose that the placement of the dragons was related to the physical behavior of water when rubbing the bowl's handles? Without the chance to engage the artifact physically – to fill it with water and rub it – why would such a conjecture ever come to mind?

It is an empirical question whether people are likely to consider an artifact to have a function that requires actions unavailable to them. But, even if an archaeologist were to entertain that conjecture, how could he or she test it unless they filled the bowl with water? Admittedly, such an interaction could be recreated in a virtual environment. But why would that feature be present? It is not part of the resource system specific to the dig. As designers of virtual environments how can we know, in advance, what may be useful to inquiry? It is hard to anticipate archeological creativity.

How do we think with objects?

Proponents of distributed, situated, embodied, embedded, enactive cognition, and proponents of extended mind – all fashionable views today – accept, in one form or another, that people think in illustrations, diagrams, mathematical symbols and language. When we engage these material things, we do not just harness them, we actually think with them. They serve as material vehicles for thought. I will not review the arguments for this position here. The arguments most readily made, typically, are for material vehicles that represent propositional thought.

Ancient artifacts, however, may mediate thought differently. They may have more to do with non-linguistic thinking. To date, there is no adequate theory explaining how humans co-opt non-propositional things for thought. Pity. Archaeologists would be well served by such a theory. If only we knew how musicians think with their musical instruments, or how a chef uses cooking utensils to think, we might better understand how Paleolithic or later peoples use blades, axes, and fire to think.

[3] http://www.interdisciplines.org/artcognition/papers/7 Drawing in the Social Sciences: Lithic Illustration by Dominic Lopes

[4] The term professional vision was introduced by Chuck Goodwin (1994).

A simple example of object mediated, non-propositional thought can be appreciated by looking at the small artifact representing a boar, shown in figure 4. Behind the boar is its '3D' sketch as it appears in a graphic program. Such graphic programs are used to recreate these artifacts in 3D virtual environments.

How would a child playing with a physical version of the 3D boar experience the boar differently than a child playing with its virtual counterpart?

Consider this. A natural game with the physical version is to animate the toy, to move it around in order to simulate running, rutting, and so forth. Why is this non-propositional thinking? Because during such play attention focuses as much on movement as on narrative. For instance, the child might animate the boar as if it is running toward a wall. Imagination is important in this process. What might the contents of this imagination be? To a physically orient child, the interesting element might be how the boar must stop and then turn around at the wall; or, how a hunter might cut off its exit and trap the animal. Clearly, there is narrative here. But, the value of the story is not so much how it can be glossed in words; its value lies in the physical lessons the child acquires through imitation and play: how to hunt, how to 'think' like a boar.

Might the same form of play take place in a virtual environment? Perhaps. It is an empirical question requiring experimental test. But, if you believe that the look and feel of the toy boar matters, then we share the same bias. It is not that we know there is something intrinsic about 2D objects making them unsuitable to serve as vehicles of thought in the right way. As long a person can interact with an object easily, whether those objects be digital structures on screens, or pencil sketches on paper, they can recruit attributes of those objects for thinking. This is especially true for representational objects such as illustrations, symbols, and so on. People interact with them in a thinking way every time they use them to solve problems. But, if it is hard to interact with these things, if it is difficult to rotate the digital boar, for instance, or if it is hard to move it forward in a way that lets one imaginatively project oneself into it's hooves, then the user's experience will be significantly less in the virtual than in physical worlds, so much so that it the user cannot think with the digital boar immersively.

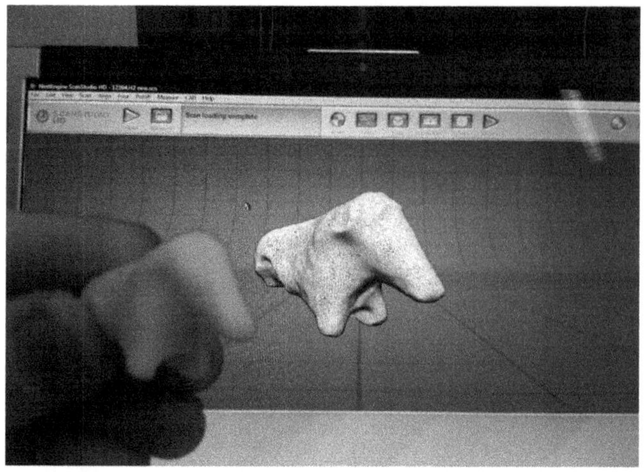

Figure 4. A 3D print of an artifact found at an archaeological dig (Maurizio ref) is held in front of the graphical program that was used to encode its dimensions and drive the 3D printer. It is easy to understand how much more we can do with the tangible 3D object than with its '3D' image using a mouse or wand.

Supporting imagination and projection

It is worth exploring this idea further. I have argued, so far, that engagement with material artifacts can enable thought, and that tangible interaction is often required to stimulate imagination of the function of artifacts. 2D representations, might in principle, serve as vehicles for thought, but it depends on how a thinker can interact with them.

An arbitrary 2D representation, such as a typed or written word, cannot literally be moved once it is written down because it has no material form. It can be re-written in a new position; it can be erased or annotated. But it cannot actually be moved unless its material support, the paper it is on, is moved. In the case of a digital screen, the digital image of word can be moved if it is grabbed by a mouse and dragged. The experience of a person with a 3D toy is quite different. People can reach out and touch toys, they can move them, pretend to make them dance or prance, and so on. How does this change the context of thought?

To explore this deep problem it is necessary to reflect on how a person can appropriate something as a vehicle of thought. Look at figure 5. Can the six pieces on the left be assembled into the target on the right?

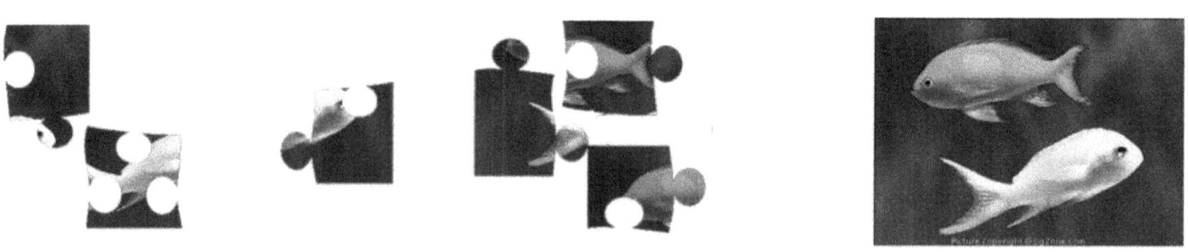

Figure 5. Can the six pieces on the right be assembled to create the picture on the left?

Figure 6. To look at this card tower is to see it as ready to topple at any moment, and most especially as ready to topple if one imagines pulling on a card. This is a form of prospective perception.

We can imagine two conditions: a physical condition where the subject can move the pieces, and physically try assembling them into the target; and a mental imagery condition, where the simulation or proof procedure is performed in the head, presumably through mental rotation and mental assembly. If the mental imagery condition is considered to be thinking, and mental images of the pieces are taken to be vehicles of thought, should we not consider the physical process of rotating and assembling the pieces in the physical condition also to be part of the thinking process? We may be unclear exactly what constitutes the vehicles of thought. For instance, are they the physical pieces, or the changing perceptions of the physical pieces, or the distributed process of moving while perceiving the movement of each piece? But in two out of the three interpretations, the vehicle contains something outside the head. People are assumed to include the pieces as part of their thinking process. Their manual control extends to the pieces, a bit like the way a blind man's perceptual system is extended to include his cane. The result is to extend the envelope of thought to include material things outside the head.

The next step is to explain how this interactive process works. It is my view, that the core interactive strategy of humans involves a back and forth process of projecting structure on the world – seeing the world as it might be after a few plausible actions – and then actualizing one or another of those possible worlds. Our projection or our capacity to imagine is anchored in the world as we perceive it, but we can augment it or partially alter it by making a few quasi-perceptual changes. I say quasi-perceptual because in perception we sometimes experience the world not just as it is at the moment but as it is dispositionally. For example, in figure 6, we do not just see a tower of cards, we see a fragile structure, ready to fall. Its disposition to fall is not a visible attribute, but a phenomenologist would say it is part of our experience of the tower. Certainly, someone who thinks they might pull on one of its lower cards 'sees' it 'as' ready to topple.

If I am right, and a core interactive strategy is to project then actualize then project more, then the trajectory of thought will be sensitive to the actions we can perform. For instance, in figure 7, our capacity to solve the geometric problem will be sensitive to how easy it is to make actual some of the constructions we consider as we reflect on the problem. If it is very hard then subjects tend to do more mental projection before actualizing. If it is easy, then they may not project much more than the line they intend to actualize. They work primarily with their pencil in hand.

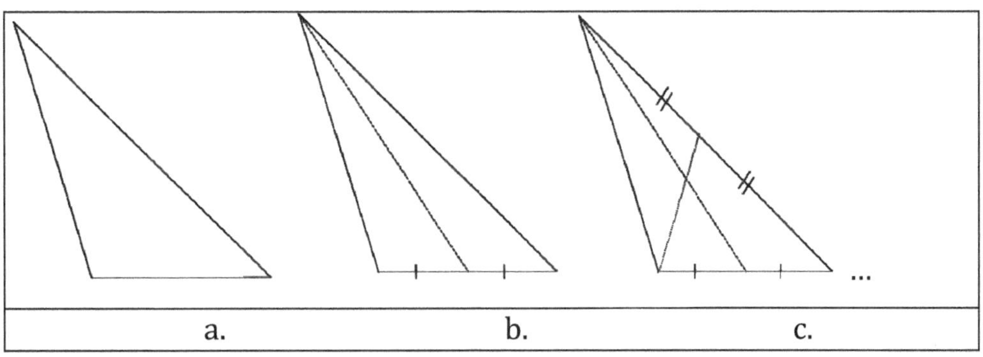

Figure 7. To solve the problem of whether the medians of an arbitrary triangle meet in a single point (the centroid) a typical subject begins by making a diagram then mentally projecting a median, as s(he) begins to think about the problem. Many subjects continue with a second or even a third projection of the median before picking up their pencil again to actualize and annotate their projections.

The point is simple: we are able to project farther into possible worlds, to see how the world might be, if we can physically play with objects. Environments where we can perform these actualizing actions easily, or at low cost, are ones where we will find it easier to think more deeply. We are able to incorporate bits of the external world into our thought processes. For virtual worlds, the challenge is to support this sort of interactive approach. Make it easy to project, then easy to actualize, and so on.

Conclusion

I have argued that cyber-archaeology offers tremendous possibility for extending the cultural and scientific experience of professional archeaologists and ordinary citizens by lowering the cost of adding digital information and visualizations to digs. But to meet the deepest needs of archaeologists, virtual environments must move beyond the purely visual to include the tangible. The need is driven by the very nature of thinking. As we improve our understanding of the way thought reaches out to include material objects we control, it is apparent that archeaologists rely on playing with the material artifacts they discover to deepen their understanding. Nothing, in principle, prevents future virtual environments from supporting this material engagement of digital versions of artifacts. But, there is much that remains to be understood about how to realize this material engagement, both at a technological and a cognitive level. The result will be worth the wait.

Acknowledgements

I thank Maurizio Forte for conversations and encouragement on this topic, and for the National Science Foundation for their support through Project #081386SX SGER: A Study of Following Instruction in Simplified Virtual Environments and Natural Environments: Evidence of Transfer and Comparisons of Cognitive Efficiency.

References

Addington, Lucile R. (1986) *Lithic Illustration: Drawing Flaked Stone Artifacts for Publication*. Chicago: University of Chicago Press

Carrello C., & M. Turvey (1996). Rotational Invariants and Dynamic Touch. In M. A. Heller (Ed.), *Touch, representation, and blindness* (pp. 27-66). Oxford: Oxford University Press.

Clark, Andy. (2003) *Natural-born cyborgs: Why minds and technologies are made to merge*. Oxford: Oxford University Press

Forte, Maurizio. "About virtual archaeology: Disorders, cognitive interactions and virtuality." In, *Virtual Reality in Archaeology.* J. A. Barceló, M. Forte, and D. H. Sanders, eds., pp. 247-59. Oxford: BAR International Series (843), 2000.

Goodwin, Charles (1994) `Professional Vision', *American Anthropologist* 96(3): 606-33

Hutchins, Edwin. (2005) Material anchors for conceptual blends. *Journal of Pragmatics*, Volume 37, Issue 10, October 2005, Pages 1555-1577

Kirsh, David. (1995)The Intelligent Use of Space. *Artificial Intelligence,* Vol. 73, Number 1-2, pp. 31-68,

Kirsh, David. (2009). Problem Solving and Situated Cognition. In Philip Robbins & M. Aydede (eds.), *The Cambridge Handbook of Situated Cognition*. Cambridge

Kirsh, David. (2010) Thinking with External Representations. AI & Society: *Knowledge, Culture and Communication,* Springer. London. (in press)

Latour, B., 1986: Visualization and cognition. Thinking with eyes and hands, in H. Kuklick, and E. Long (eds), *Knowledge and Society. Studies in the Sociology of Culture Past and Present* Vol. 6, 1-40.

Lopes Dominic: http://www.interdisciplines.org/artcognition/papers/7 Drawing in the Social Sciences: Lithic Illustration

An open source approach to cultural heritage: Nu.M.E. Project and the virtual reconstruction of Bologna

Nicola Lercari
University of Bologna, Italy

Abstract

Nu.M.E. (New Electronic Museum of the city of Bologna) is a multidisciplinary project involving historiography, museum studies, and computer science. Since the late 1990s it has been answering complex methodological questions with the aim to concentrate, on a single digital platform, the results of many years of research in the field of urban history. The case study of Nu.M.E. is the city of Bologna, in Italy, and its millenary tradition of urbanization. In its 2009 release, the project strives for the definition of new methodologies of cultural transmission for the general public. Nu.M.E. 2009 also adopts a new perspective based on cutting edge, open source technologies. Such choices make it possible to demonstrate that 3D graphics, real-time engines, and interactive storytelling can be useful tools for philological representations of ancient urban landscapes. At the same time, our research evidences that these technologies are fundamental resources for contemporary museology. This paper surveys the problems that emerge in analysis and communication of the medieval urban landscape of thirteenth-century Bologna and will present a new method of interpretation, and visualization of historical data.

Keywords: comparative analysis, sources-based virtual reconstruction, open source, interactive storytelling.

Introduction

The conformation of Bologna's urban landscape is the result of a very long tradition of human activities. Since the Iron Age, when the small village of Felsina was founded by Etruscans in that area, *c.* 510 BC (Sassatelli 2005), the site of current city has been involved in several phases of historical changes. During the Roman Empire, Middle Ages, Renaissance, and Modern Period many social, political and economic phenomena played a role in the definition of the city's space. This fertile historical context has generated a vast, and composite amount of sources and data, most of which are still today available in local archives and libraries.

The Nu.M.E. (New Electronic Museum) project was started in the late 1990s by Francesca Bocchi, professor of Urban History at the University of Bologna. It is an attempt to define a valid methodology for studying Bologna's urban history using a spatial approach and digital technologies. As a decennial research project, Nu.M.E. has over time adapted to the evolution of computer graphics (Diamanti 2007). It has faced, in every single version, different historiographical issues and technical challenges (Guidazzoli and Bonfigli 1999). The original idea of adopting a rigorous and philological attitude toward virtual reconstruction has never changed, though. During the past decade Nu.M.E. has dealt with the three-dimensional reconstruction of some parts of the historic centre of Bologna with the aim to visualize transformations of urban structure and road networks, from the late Middle Ages to the present.

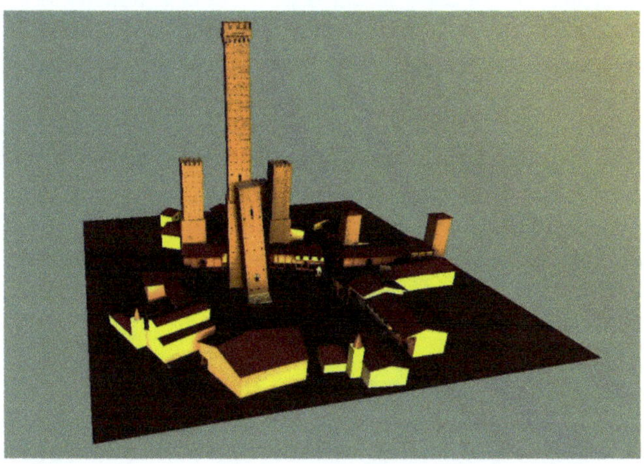

Figure 1. Part of thirteenth-century Bologna urban landscape around *Piazza di Porta Ravegnana*.

In 2009 new emphasis is given to Nu.M.E.; this happens because some local stake holders, such as municipality of Bologna and *Fondazione Del Monte*, decided to support the creation of the Museum of the City of Bologna (which will open to the public, approximately in 2011 in the medieval complex of *Palazzo Pepoli Vecchio*). The circumstance introduces a new moment of reflection on cultural transmission techniques for the general public. It also opens the way for the beginning of our research, about the communicative aspects of virtual heritage projects.

Nu.M.E. 2009 historical context

This release of the project focuses on one of the most representative moments of Bologna's urban history: the thirteenth century . According to Bocchi (2007, 233-241), the death of the Holy Roman Emperor Henry VI, in AD 1197, determined Bologna's independence and had very important outcomes on the city's destiny. The municipality obtained absolute power, and was able to concentrate a great amount of economic resources in the creation of public infrastructures. In AD 1200 a relatively

Figure 2. Thirteenth-century porches.

Figure 4. View of the no longer existing *Artenisi* and *Guidozagni* towers.

Figure 3. Thirteenth-century buildings.

long period of peace began, and an ambitious project of urban development started (Huber 2000, 3-9). It was characterized by the construction, in a temporal range of about a century, of the main urban public space (called *curia Comunis*, which corresponds to current *Piazza Maggiore*) and several new buildings (*Palazzo del Comune, Torre dell'Arengo, Palazzo di Re Enzo, Palazzo della Biada*) with political and economic functions, all around it. A wide suburban area (*Campo del Mercato*) was dedicated to the weekly cattle market, while an extensive system of canals was dug to the benefit of river navigation and textile manufacturing. To defend the city from a potential enemy attack, the construction of an impressive fortification (the so called *Circla*) was started, as well. The date of its completion was AD 1327, but it is very probable that in the late 1220s the whole city was already surrounded by a moat and a palisade (Bocchi 2005, 243-246).

For these reasons, the thirteenth century can be easily seen as a context of a radical process of spatial reconfiguration that completely modified the late medieval structure of Bologna. These historical dynamics shaped the central part of the city so intensely that they even created the basis for the current configuration of downtown. All the mentioned historical events are furthermore documented by an abundant set of both iconographic and documentary sources. Such characteristics make late medieval Bologna an optimal context for doing research on virtual reconstruction techniques; they also give us the opportunity to test different methodological and communicative approaches on a solid informative basis.

Sources-based virtual reconstruction

While other disciplines that deal with the interpretation of the past and relics take advantage of spatial relations interpretation, the study of history, rarely, has the opportunity to benefit from spatial analysis of data. With this in mind, we reflect on the use of digital technologies in historical research and try to answer the following questions: is the comparative analysis of ancient sources a valid methodology for the virtual reconstruction of early cities? Can iconographic material be a useful resource for the development of a virtual museum on Middle Ages lifestyle in Bologna?

The development of Nu.M.E. 2009 helps us find answers. This release of the project demonstrates that 3D graphics and real-time engines can be useful tools for a philological representation of an ancient urban landscape. Working on three-dimensional scientific visualization of a set of historical information, we determined that a traditional historiographical description of an urban place is not sufficient to be used as base of modelling activity **(Put Figure 5 here)**. The geometrical representation of historical places, that time has altered or destroyed, needs a very deep knowledge of all urban structural components. This circumstance pushes historians to find new relationships between data and sources, in the attempt of gathering very precise information on both architectural elements and construction techniques. At the same time, it requires that 3D modelers be dialectically involved in a discussion on content creation. It is also necessary that computer scientists, while working on visualization of the models, be aware of all the problems and needs belonging to historiographical investigation. The result of the methodology that we used in Nu.M.E. is a multidisciplinary approach that opens new perspective for urban history, and reduces the gap between humanities and information science.

Figure 5. Detail of the no longer existing *Riccadonna* tower.

Figure 6. East view of Via Emilia.

Let us take a step back and focus on the description of the specifics of our case study.

In Nu.M.E. 2009, the spatial visualization is focused on the most important parts of 13th century Bologna, specifically the area between current *Piazza di Porta Ravegnana* and *Piazza Maggiore* **(Put Figure 6 here)**.

The historiographical survey of this area is based on comparative analysis of documentary and iconographic sources. All the information is recorded in relational databases and, then, connected to a Geographical Information System (Bocchi 1999b, 11-28). This method permits both the development of spatial frameworks for analysing human activities on the urban landscape, and the implementation of a robust grid for virtual reconstruction. To get a scientific value this methodology follows guidelines and criteria about the study of urban history, issued by The International Commission for the History of Towns (ICHT).

Comparative analysis, in the study of Bologna's past, is a condition that can be effective because the history of the city is evidenced by a very significant documentation; that is because a great amount of historical material has been produced by notaries, religious institutions, and municipality officers since the late Middle Ages. Such a rich data set is constituted of several private and public documents:

- rental contracts of lands between religious institutions and private citizens (*enfiteusi*);
- real estate property contracts;
- medieval cadastre (*Estimi*) held by the municipality since the second half of 12th century;
- early urban planning documents (*Libri terminorum*) that described, in a very precise way, the position of pillars, porches, balconies, and roofs, in order to define the limit between public and private space, both on the ground level, and in elevation.

The work of prestigious Italian artists and cartographers, furthermore, gives our three-dimensional representation of Bologna the opportunity to be founded on several iconographic sources. The visual material we used in Nu.M.E. can be subdivided in two different sections. The first consists of a composite set of early matter such as:

- miniatures (e.g. Drappieri Miniature – AD 1523 - Giovan Battista Cavalletto – Civic Medieval Museum, Bologna, Italy);
- frescos situated in Bologna (e.g. *Madonna del terremoto* – AD 1505 – Francesco Francia – *Palazzo Comunale*, Bologna, Italy);
- frescos situated in other cities (e.g. Allegory of Good Government – Ambrogio Lorenzetti – 14th century – *Palazzo Pubblico*, Siena, Italy);
- perspective Renaissance cartography (e.g. The perspective map of Bologna – Scipione Dattili – AD 1575 – The Pope's apartments, Vatican City);
- topographic maps (e.g. Topographic map of Bologna – AD 1833).

Figure 7. Detail of *termini* and porches in *Piazza di Porta Ravegnana*.

The second section, instead, is made of recent images obtained through optical instruments:

- early 20th-century pictures collections (e.g. *Collezioni d'Arte e Storia della Cassa di Risparmio in Bologna);*
- aerial photos;
- satellite photos.

Thanks to this great availability of historical sources, the accuracy of 3D models in Nu.M.E. is very high.

This goal has been obtained by performing complex methodological tasks such as:

- quantitative analysis of cadastres and legal acts;
- architectonic analysis of building components (e.g. pillars, windows, doors, etc.);
- spatial surveys, about volumes and arrangement, of projecting roofs, balconies, porches and other above ground assets;
- reflections on physical and mechanical properties of materials and buildings;
- ancient cartography analysis;
- comparison between early and recent representations of the city;
- analysis of monumental complexes and geographical landmarks;
- systemic comprehension of urban structures and road network through Geographical Information Systems.

Such a rigorous approach permits us to get a correct and precise virtual reconstruction of the urban space of Bologna. It also defines a standard for representation of early cities through digital technologies. Although it is possible to choose different forms of virtual reconstruction, it has been demonstrated that our methodology can be considered as an effective way to keep a virtual heritage project in balance between technological dimension, communicative effectiveness and cultural value.

Open source technologies for virtual heritage

The originality of our research project is not only methodological. The technological perspective of Nu.M.E.

Figure 8. Spherical Harmonics Lighting technique in OGRE 3D graphics engine.

2009 brings innovation in the field of cultural virtual environments and can be seen as a mature example of 'musealization'. We developed a workflow that starts with historiographical investigation on medieval sources, designs a specific communicative process based on them, and eventually implements a complex visualization system based on cutting edge technologies. This is because we decided to focus our research activity not only on the development of historiographical contents for the virtual reconstruction, but also in 3D visualization. In computer science the term 'open source' indicates a software released with a type of license that permits the source code to be available to anyone who requests it. In this way developers can take advantage of the collaboration, usually free and spontaneous, of a community of users. The final product of this collaborative process can reach a level of complexity and efficiency that could not be obtained by a single group of programmers.

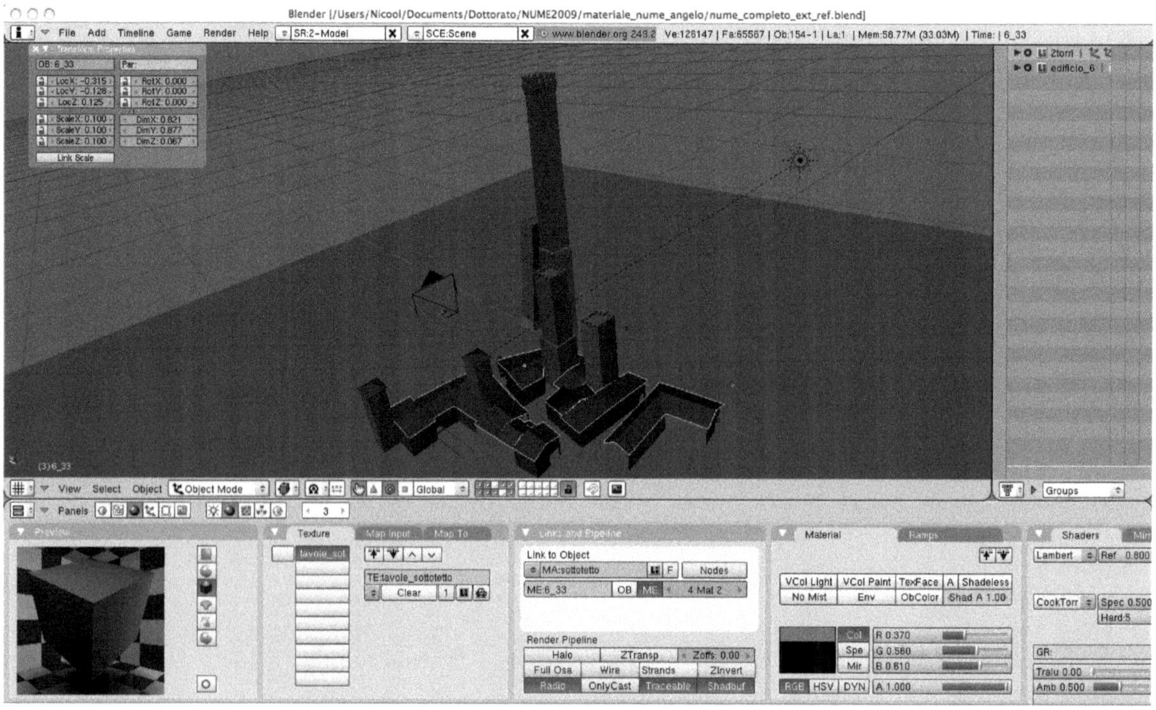

Figure 9. Low polygon Bologna 3D model in Blender.

The model of development we adopted in our project can be defined as open source, as well. The approach of Nu.M.E. 2009 can be described in this way because it pursues a more ethical use of technology. The need to identify effective and alternative models of development is a priority in a field, such as virtual heritage, in which the availability of funding is often limited. An open source approach is particularly effective because it is aimed at obtaining a collective cultural progress based on sharing of expertise between groups of developers and users. This more ethical use of technologies allows for evident advantages such as:

- reduction of software licensing cost;
- sustainability of projects;
- sharing resources and results with other academic institutions;
- extreme portability of file formats;
- quality and usability standards based on community review.

The technological aspect of Nu.M.E. 2009 can be seen as an example of an open source approach in the field of virtual heritage. What follow is a brief overview of all the technical components that are involved in our project. All computers that we used to implement and visualize the virtual environment were running the Unix-like operating system Gnu/Linux Ubuntu. 3D modelling and UV unwrapping texturing have been performed using the powerful open source graphics application Blender. The activity of texture editing and retouching has been achieved using the free software raster graphic editor Gnu Image Manipulator (GIMP). Three-dimensional visualization of thirteenth-century Bologna takes advantage of the powerful rendering engine OGRE (*Object-Oriented Graphics Rendering Engine*).

Specifically, OGRE is a C++ scene-oriented engine that allows for the development of interactive applications based on hardware-accelerated 3D graphics. Rigid body simulation in Nu.M.E. 2009 is possible thanks to Bullet physics library. Bullet is a open source library that is

Figure 10. Bump mapping technique in Ogre 3D graphics engine.

Figure 11. Virtual characters in OGRE 3D graphics engine.

considered as a standard for video games and film post-production in entertainment industry. It uses multi-threaded 3D collision detection and soft body, and rigid body dynamic libraries. To implement realistic shading effects we designed a Spherical Harmonics Lighting system running on CUDA, the cutting edge parallel computing architecture of Nvidia; this powerful tool permitted us to pre-process all the effects of light directly on special images called light maps, and then display them in real-time using a Nvidia GTX 280 graphic card. In virtual navigation of Bologna, passage of time is simulated using OpenGL shading language and a special module of OGRE called CAELUM. Video capturing for presenting our project to conferences and showing the result of our work online is based on Yukon OpenGL capturing framework. This is a set of libraries and applications that are designed to capture realtime videos of OpenGL applications with high frame rate.

The introduction of every new technology implies an initial moment of uncertainty before giving effective results; our choice to base Nu.M.E. on open source technologies demonstrated that even these new digital tools have pros and cons. In the first phase of the project we encountered some major technical problems and certain difficulties in handling such experimental technologies. Despite those issues, the good quality of the virtual reconstruction and the excellent results in visualization demonstrate that our choice was brilliant. If we analyse the technological quality of our work on the basis of the cultural and economical goals obtained, we can easily say that our open source approach in virtual reconstruction of early cities can become a good example for forthcoming educational and research projects in this field.

Cultural models for communication to the general public

The main focus of Nu.M.E. 2009 is the implementation of an efficient communicative process for the virtual museum of the city of Bologna. It has been stated (Antinucci 2007, 5-25) that fundamental components of cultural transmission in museums are comprehension and motivation. It has also been proved that digital technologies permit involvement of cognitive and emotional factors in the process of improvement of museum communication (Monaci 2005). In Nu.M.E. 2009 we try to follow these important principles of museum studies with the aim of helping the general public know the history of towns. This attempt pushed us to confront several theoretical and methodological problems. How can complex historiographical data be communicated to the visitors of a virtual museum of the city? What new models of interaction can improve cultural reception in a real-time simulation environment?

According to Vitali (2000) virtual reconstruction technologies can be defined as 'history commutators' because they invert the regular flow of information within a historical communication system. This condition occurs because, rather than transmitting information through a linear path, 'history commutators' attract people inside the simulated historical context, and involve them in a synesthetic process of communication in which both tangible and intangible elements of the past can be transmitted. The novel informative system that characterized 'history commutators' can be defined as reticular; this because, within this new digital context, the process of meaning creation is based on one side on a net of relations between visualized elements and on the other on a matrix of differences between those components and users. Such a new cultural model allows for communicating to the general public historical memories of specific events or human activities. At the same time spatial relations between places that no longer exist can be included in this process of three-dimensional simulation. Thus, causality and spatiality become fundamental factors of cultural virtual environments aimed to communicate historical changes. With this in mind we started to reflect on communicative codes to use in our project. From a communicational perspective, Nu.M.E., being a multi-sensorial digital environment of simulation, can be considered as media, specifically new media. Therefore the comparison of Nu.M.E.'s language with patterns and paradigms of other media is a crucial point of this research about cultural communication in virtual environments. What is the most consistent, tested, and effective cultural model of audio-visual simulation? The first answer we give is cinema. The centenarian tradition of audiovisual simulation that characterized this medium has already become useful matter for new media. It has been demonstrated that cinematic conventions such camera, screen and spatial editing are cultural interfaces in the field of digital communication (Manovich 2001).

Cinematic formal structures are essentially based on the representation of narratives within time and space. In cinema symbolic systems determine an emotional involvement of the audience; the outcome of this phenomenon is high-motivated spectators who try to comprehend transmitted information. Cinema is not the only cultural model we can take as an example. We want to include in our analysis another contemporary communicational model based on 3D graphics and real-time visualization. In the scenario of computer-based communication video games are one of the few examples of 'native' cultural paradigms (Manovich 2003, 13). The relationship between interactivity and narrativity play a creative role in video games. This very popular form of digital communication uses algorithms to customize in real-time the virtual experience of users. They shape narrative sequences, virtual characters, and virtual worlds according to the evolution of the game. At this point, we try to verify if these cultural models are useful for Nu.M.E. communication. The hypothesis is to transform virtual space of thirteenth-century Bologna in intriguing interactive adventure through historical information. First of all, we have to analyse the semantic role that narratives play in the process of information comprehension within virtual environments. Are the basic functions of any narrative structure, such as causal system, time and space, elements that can be integrated into the interactive structure of cultural virtual environments?

Causality, in a virtual environment, means possibility of constructing a specific exploration path through the composite matrix of information that is recorded in historical databases. Our objective is to determine if users of virtual museums can be guided along the process of the creation of historical meaning. We can achieve this goal using communicative techniques able to focus users' attention on systemic comprehension of the relationships that exist between 3D objects. The consequence of this improved learning process is an enhanced comprehension of historical data. We think this makes it possible to provide the general public with a deeper knowledge of cultural contents. The causal system can, furthermore, be used to mitigate the impact of virtual experience, especially when it is based on immersive virtual environments. We consider such a condition as a necessary method to avoid a cognitive overload in the audience of these complex simulation systems. Narrative can also be considered as an attempt to guide users' cognitive efforts along the main goal of our virtual museum: the allocation of a specific historical meaning to the components of reconstructed virtual space. In the summer 2009 we tested our urban simulation system with the general public and noticed that the possibility of a progressive lack of interest is very high if users are not guided along themed pathways; if this undesirable circumstance happens, visitors of virtual museum, immediately, decide to stop their cultural fruition, causing the communicative system to fail. Analysing the concept of space in virtual environments it is easy to verify that 3D simulation systems attribute many different meanings to spatial relations. Space is the interface through which users communicate with the system. At the same time, it is the medium that allows communication of information between people involved, simultaneously, in the process of simulation. Specifically virtual space, as a navigable space, is a key element in our perspective. It is the digital framework in which we can enact interactive storytelling. It becomes a path to follow, during this type of historical simulation. Users of interactive storytelling systems can express their subjectivity dealing with the virtual space (Blom and Beckhaus 2002). They can also be emotionally involved in the narrative structure describing events of the past. Thus narrativity, giving to virtual spaces a new psychological value, enriches the use of 3D models with a new emotional dimension. Spatial structure can, therefore, be exploited to give a specific shape to navigation. This can be done with specific stylistic choices that transform the concept of space; from an initial state of mere digital information virtual space becomes an emotional resource. Emotions influence users' expectations and motivation, and therefore, historical comprehension as well. Narrative structure has another important feature that can enhance the simulation of historical changes. Interactive storytelling introduces in the virtual world a temporal variable. Such a new feature determines the end of the hegemony of the spatial dimension in the production of cultural meaning within virtual reconstruction. Temporality, in cultural virtual environments, can be considered a convention that establishes chronological progression of the simulated historical events. The indefinite time of virtual worlds can, eventually, be declined in functional structures that help to enrich user experience. The outcome of this new feature is that temporality is able to describe how historical places have been transformed over time.

Using a narrative structure in cultural virtual environments creates the opportunity to intervene on simulation in order to generate a specific chronological meaning in represented historical events. We believe that the relationship between the opposite formal structures of interactivity and narrativity can open new communicative scenarios in the field of virtual heritage.

After this theoretical preamble, we want to describe, in brief, the attempt of enacting interactive storytelling in Nu.M.E. 2009. Our integration of the language of cultural virtual environments with the communicative codes of video games and cinematic language is pursued in the creation of themed pathways based on historical information. The final goal is to help users handle the interactive environment of the virtual museum of Bologna and decode complex historical information. Such decoding takes advantage of the simplification that storytelling and spatial-temporal structures give to virtual reconstruction. Although they can be seen as denigratory features in virtual environments aimed at historical interpretation, we strongly believe that these formal structures are not a limit. Interactive storytelling enriches the communication of cultural heritage to the general public of museums, or virtual museums. This happens because the 'narrative-interactive model' stimulating users' emotional spheres and, helps their cognitive system in the learning process of represented historical data. This is a key point in our research project about Nu.M.E.'s communicative aspects. Even though user experience of the simulated medieval space of Bologna is, primarily, a matter of perception and interaction, we try in the 2009 version of the project to describe scenes of medieval daily life or historical memories of important events that characterized thirteenth-century Bologna. Our effort is focused on the representation of a cloth market of that was held in *Piazza di Porta Ravegnana* since the late Middle Ages. This economic activity was so relevant for the city that the municipality assigned to it the square just below the Garisenda and Asinelli Tower; this central location is one of the most significant and symbolic places of mediaeval Bologna, (De Angelis 2000). Thanks to information we can get from the fifteenth-century *Drappieri Miniature* is possible for us to create an accurate themed pathway between market stalls, and involve users, especially young students, in a stimulating process of comprehension of production activities that characterized the Middle Ages.

Figure 13. View of *Cappella della Croce* in *Piazza di Porta Ravegnana*.

Figure 12. Scene of medieval daily life in *Piazza di Porta Ravegnana*.

Figure 14. View of *Garisenda*, *Asinelli*, and *Artenisi* towers.

Conclusions

Technological evolution and new cultural paradigms push museology to find different models for museums of the future. Just five years ago the majority of virtual museums were only websites in which users could just experience a digitized version of artwork and collections. Today's cultural communication can take advantage of several new methodologies and tools that, in the recent past, were confined to 'techno-scientific imaginary' of sci-fi movies. 'Multiuser online metaverses', collaborative virtual environments, and embodied learning processes open new horizons for contemporary museology (Forte and Bonini 2008). Nu.M.E. challenges virtual heritage methodology in the attempt to contribute to this cultural debate with a decennial tradition of research and innovation. In the 2009 release of the project we try to give an example of sustainable communication processes that might become useful for forthcoming initiatives. The purpose of our research is to test open source technologies for the communication of historical contents. In this way we try to define a novel paradigm for museums which want to use 3D graphics and simulation technologies. We reflect also on interactive storytelling in order to introduce new formal structures in the communicative context of Nu.M.E. Our final aim is involve visitors of the virtual museum of Bologna in a emotional and intriguing experience of the past of the city. The outcomes of these new features are the possibility to aim our virtual reconstruction to the general public. The use of reconstructed historical context in early versions of Nu.ME consisted of 3D navigation in the first person where users could just use 3D objects to perform simple actions. The introduction of a narrative structure enriches this system of cultural communication. It gives the user an active role in the process of the creation of meaning based on the simulation of significant scenes and events of late Middle Ages life. The great realism and historical accuracy of 3D models allow for obtaining scientific validation and enlarge our project perspectives and goals. Nu.M.E. 2009 becomes a useful tool for both historians and general public. Experts of urban history can take advantage of our virtual reconstruction in the interpretation of spatial-temporal relations and attribute a precise meaning to the urban landscape. Tourists, youth, and students can find in Nu.M.E. 2009 a non-conventional, intriguing, but at the same time historically correct, way to get in touch with the past of Bologna and comprehend its long history.

The final result of our methodology can be, therefore, defined as a 'sources-based virtual reconstruction' of thirteenth-century Bologna that is both philological and emotional. Thus, we can finally say that our goals of improving communicative aspects of Nu.M.E. and keeping decennial scientific and cultural value of the project have been achieved. We are, also, aware that the relationship between cultural heritage and technology is today related to their social dimension. A contemporary virtual museum can not exclude dealing with what is radically changing user experience and new media communication. Therefore, we considerate Nu.M.E. 2009 not the end of our investigation on communicative aspects of virtual museums, but the starting point for a new release of the project focused on the collaborative aspects of cultural reception. It will be released approximately at the beginning of 2011, and will conclude this research project about thirteenth-century Bologna, adding educational, social and communicative values to virtual reconstruction.

References

Antinucci, F. 2004b. *Comunicare nel museo*. Milano, Percorsi.

Antinucci, F. 2007. *Musei virtuali*. Bari, Laterza.

Blom, K. and Beckhaus, Steffi 2005. Emotional Storytelling. *Proceedings of IEEE VR2005*. Bonn, Shaker Verlag.

Bocchi, F. 1995. *Il Duecento*. Bologna, Edizioni Grafis.

Bocchi, F. (ed.) 1999a. *Atlante multimediale di Bologna. La storia, i luoghi, le persone. Cronologia e bibliografia*. Bologna, Edizioni Grafis.

Bocchi, F. 1999b. Nuove metodologie per la storia delle città: la città in quattro dimensioni. *Medieval Metropolises, Proceedings of the Congress of Atlas Working Group*, 11-28.

Bocchi, F. Smurra R., Ghizzoni M. 1999. The 4D virtual museum of the city of Bologna, Italy. *Proceedings of ACM SIGGRAPH 1999*.

Bocchi, F. 2007. Lo sviluppo urbanistico. In Capitani, O. (ed.) 2007. *Storia di Bologna 2. Bologna nel* Medioevo. 187-309. Bologna, Bononia University Press.

Bocchi, F. 2008. *Bologna nei secoli IV-XIV. Mille anni di storia urbanistica di una metropoli* medievale. Bologna, Bononia University Press.

Bolter, J. D. and Grusin, R. 1999. *Remediation. Understanding New Media*. Cambridge, The MIT Press.

Cameron F. and Kenderdine, S. (eds.) 2010. *Theorizing Digital Culture. A Critical Discourse*. Cambridge, The MIT Press.

De Angelis, C. 2000. Bologna alla fine del Duecento: il modello ricostruttivo. In Mediaca, M. (ed) 2000. *Forme e colori del Medioevo a Bologna*. Venezia, Marsilio Editori.

Diamanti T. 2007. *Programmazione per la fruizione del progetto Nu.M.E. attraverso Internet*. Unpublished PhD thesis, University of Bologna.

Forte, M. 2007. *Proceedings of The International Conference Using New Technologies to Explore Cultural Heritage*.

Forte, M. and Bonini, E. 2008. Embodiment and enaction: a theoretical overview for cybercommunities. *Proceedings of 14th International Conference on Virtual Systems and Multimedia VSMM 2008*.

Grigoletto, F. 2006. *Videogiochi e cinema. Interattivitá, temporalitá, tecniche narrative e modalitá di fruizione*. Bologna, CLUEB.

Guidazzoli, A. and Bonfigli, M.E. 1999. The creation of the Nu.M.E. project. *Proceedings of ACM SIGGRAPH 1999*.

Harrigan, P. and Wardrip-Fruin, N. (eds.) 2009. *Third Person. Authoring and exploring vast narratives*. Cambridge, The MIT Press.

Hein, H. S. 2000. *The Museum in Transition. A Philosophical Perspective*. Washington, Smithsonian Institution Press.

Hess, R (ed.) 2007. *The Essential Blender: Guide to 3D Creation with the Open Source Suite Blender*. San Francisco, No Starch Press.

Huber, H. W. 2000. Architettura e urbanistica nel Duecento a Bologna. In Mediaca, M. (ed) 2000. *Forme e colori del Medioevo a Bologna*. Venezia, Marsilio Editori.

Junker, G. 2006. *Pro Ogre 3D Programming*. New York, Apress Inc.

Macluhan, M. 1964. *Understanding Media: The Extensions of Man*. New York, McGraw-Hill.

McClean, S. T. 2008. *Digital Storytelling. The narrative power of visual effects in film*. Cambridge, The MIT Press.

Manovich, L. 2001. *The Language of New* Media. Cambridge, The MIT Press.

Manovich, L. 2003. New Media from Borges to HTML. In Wardrip-Fruin, N. and Montfort, N. (eds) 2003. *New Media Readers*. Cambridge, The MIT Press.

Monaci, S. 2005. *Il futuro nel museo. Come i nuovi media cambiano l'esperienza del pubblico*. Milano, Guerini e Associati.

Murray, J. H. 1998. *Hamlet and the Holodeck*. Cambridge, The MIT Press.

Peck, A 2006. *Beginning GIMP: From Novice to Professional*. New York, Apress Inc.

Qvortrup, L. (ed) 2002. *Virtual Space: spatiality in virtual inhabited 3D* worlds. London, Springer.

Riva, G. 2008. *Psicologia dei nuovi media*. Bologna, Il Mulino.

Sassatelli, G. 2005. Bologna etrusca. In Sassatelli, G. and Donati, A. (eds.) 2005. *Storia di Bologna 1. Bologna* nell'antichità, 117-338. Bologna, Bononia University Press.

Sassatelli, G., Morigi Govi, C., Ortalli, J. and Bocchi, F. 1996. *Da Felsina a Bononia. Dalle origini al XII secolo*. Bologna, Edizioni Grafis.

Sparacino, F. 2004. Scenographies of the Past and Museums of the Future: From the Wunderkammer to Body-Driven Interactive Narrative Spaces. *Proceedings of ACM Multimedia 2004*.

Vitali, S. 2000. *Proceedings of conference Archivi e voci d'autorità. Metodologie ed esperienze a confronto per i beni archivistici, librari e storico – artistici*.

On-Site Digital Archaeology 3.0 and Cyber-Archaeology: Into the Future of the Past – New Developments, Delivery and the Creation of a Data Avalanche

Thomas E. Levy, Vid Petrovic, Thomas Wypych, Aaron Gidding, Kyle Knabb, David Hernandez, Neil G. Smith, Jürgen P. Schlulz, Stephen H. Savage, Falko Kuester, Erez Ben-Yosef, Connor Buitenhuys, Casey Jane Barrett, Mohammad Najjar, Thomas DeFanti

University of California, San Diego, Center of Interdisciplinary Science for Art, Architecture, and Archaeology, California Institute of Telecommunications and Information Technology, U.S.A.

Corresponding author: tlevy@ucsd.edu

Abstract

Over the past 3 years, with the establishment of the new Center of Interdisciplinary Science for Art, Architecture and Archaeology (CISA3) at UCSD's California Institute of Telecommunication and Information Technology (Calit2), a collaboratory framework has been established facilitating joint research between archaeologists, computer scientists and engineers. We report here on a cyber-archaeology field recording system that feeds into a cyberinfrastructure delivered over the Mediterranean Archaeology Network (MedArchNet) on a Google Earth platform. A field test of the new system was carried out in 2009 at Khirbat en-Nahas (KEN), an Iron Age (ca. 1200 – 900 BCE) copper production center in Jordan.

Key words: Cyber- Archaeology, LiDAR, Balloon Photography, Artifact Informatics, Cyber-infrastructure

Introduction

In 1999, when we made a commitment to 'go digital' to record all our field measurements on excavations in Jordan related to the role of ancient metallurgy on social evolution, we had no idea that our 20th c. data would be 'pre-adapted' to the growing field of 3D visualization. This first application may be referred to as on-site digital archaeology (OSDA) 1.0(Levy et al. 2001). Being based in far away San Diego with only periodic access to the field and artifact collections left behind in the Middle East at the end of each excavation season; we wanted to develop a recording system that would enable us to take our entire dataset home with us for analyses. In this case, as Plato (ca. 427 BC – 347 BC) wrote in *The Republic*, necessity was indeed the 'mother of invention' and the driving force for abandoning the old analogue paper recording system we had used for over 25 years (Levy 1987). Over the years, as computers have become more portable and more powerful, OSDA 2.0 emerged, that like Ver. 1.0, has Geographic Information Systems (GIS) at its nexus (Levy and Smith 2007) to facilitate the spatial analyses of archaeological data.

Here we summarize the most important new developments in OSDA 3.0 that make it a much more versatile system (Figure 1) that takes advantage of both off-the-shelf technologies and also includes new computer programs and hardware developed specifically to solve archaeological/cultural heritage problems that face researchers working around the world today. As a field science, archaeology depends on precisely documenting the x, y and z coordinates of excavation and cultural heritage data. By acquiring this kind of metadata for material culture, it is possible to thread together an array of different kinds of spatial and analytical data recorded in the field that ultimately relates to the larger theoretical and historical questions that drive interest in world cultural heritage.

OSDA begins with the mapping of all realms of material culture or any spatial information of relevance to the archaeological research. For archaeological field survey work, we depend on differential GPS, whereas excavation work depends on Total Stations. Both methods have a high degree of accuracy, typically around 1-2 centimeters. All artifacts, features and topographic points are stored in a GIS database that can be quickly and easily accessed for analysis or transferred to a variety of 3D computer modeling programs. Many of the other technologies we use rely on the data taken with the total station. For example, most of the imaging techniques we use are geo-referenced using control points taken with the total station. Balloon photographs, described below, are imported into ArcGIS and geo-referenced. This allows us to create detailed site plans of architecture and other features. For a more detailed description of how Total Stations and GPS are used in the field readers are referred to Levy and Smith (2007). Figure 2 shows how geo-referenced aerial photographs taken with the new helium balloon system are used for creating publishable maps at KEN. These data are subsequently used in the 3D visualization environments discussed below.

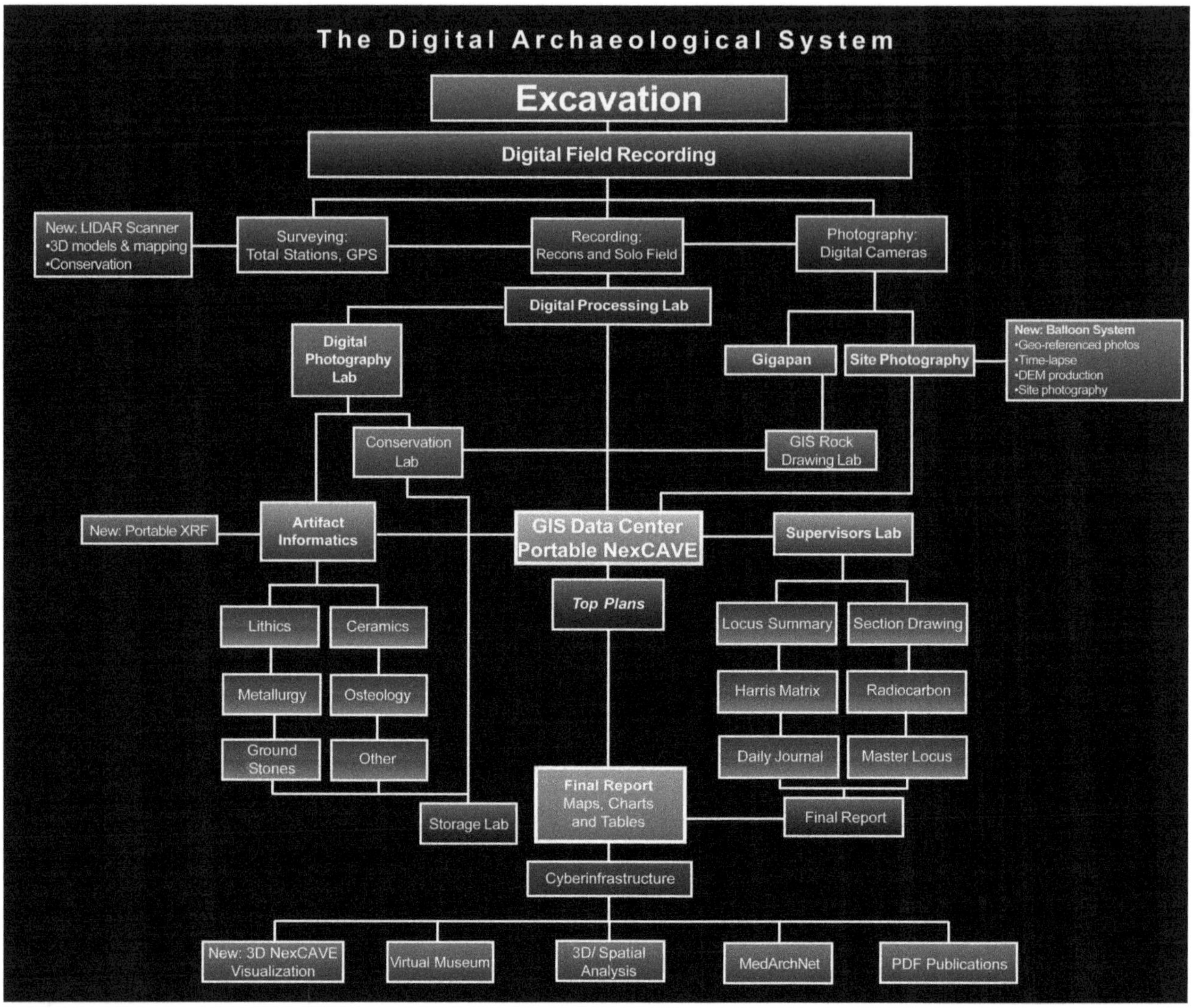

Figure 1. Flow-chart illustrating the On-Site Digital Archaeology 3.0 system with new elements highlighted and discussed in this paper: LiDAR mapping, helium balloon airborne photography, StarCAVE, NexCAVE, Artifact Informatics, and cyber-archaeology represented by the Mediterranean Archaeology Network (*MedArchNet*).

Figure 2. An image captured by the UCSD balloon system is geo-referenced and the rocks are 'traced' to create a detailed architectural plan. Shown here is part of a building complex from Area W, KEN viewed in ArcMap.

Making more out of X, Y, and Z: New Imaging Techniques Used in OSDA 3.0

The new OSDA 3.0 system integrates a number of imaging techniques including a specially designed helium balloon platform for taking vertical and oblique photographs of archaeological/cultural heritage sites, GigaPan photography that enables multi-gigapixel panoramas to be taken of sites with DSLR cameras, and terrestrial LiDAR.

a) Airborne Balloon-Based Imaging

Surface level photography provides a record of excavations, however to place the site in its more general context for larger scale analyses, aerial photography is needed. Here we describe improvements on the 'boom' system used in OSDA 1.0 and 2.0. The early method involved placing a small digital camera on a ca. 7 m long wooden stick, struggling to lift it and capturing an image covering only a ca. 5x5 m excavation square (Levy and Smith 2007). This was time consuming, awkward and involved long hours of 'stitching' geo-referenced photos together to cover one excavation area. These problems were a catalyst to design our airborne helium balloon system in the summer of 2009 by UCSD CISA3/Calit2 undergraduate students that was deployed in Jordan in September of that year. It consists of a helium-filled balloon with a sail appendage mounted on a stable aluminum platform equipped with two 15 megapixel digital SLR cameras monitored with a live-feed (Figure 3). The balloon (Kingfisher™ Aerostat from Southern Balloon Works) is tethered to an operator on the ground that can position it over a specified excavation and cover areas up to 700 m^2. The system can capture images up to a height of ca. 200 meters at high resolution and excellent stability that insure image quality. The system is versatile enough that it can be brought down to a lower height for even higher resolution image capturing.

The vertical high-resolution images of site architectural features are geo-referenced and aid in the creation of publication quality maps using ArcMAP in the ArcGIS 9 suite. As all photographs are shot as stereo-pairs that canbe used to create digital elevation models (DEMs). For oblique views, the cameras can be manually rigged to supplement more interpretive analyses.

Another requirement for this system was quick deployment on a daily basis. Thus complicated rig systems were avoided in the design of the aluminum platform.

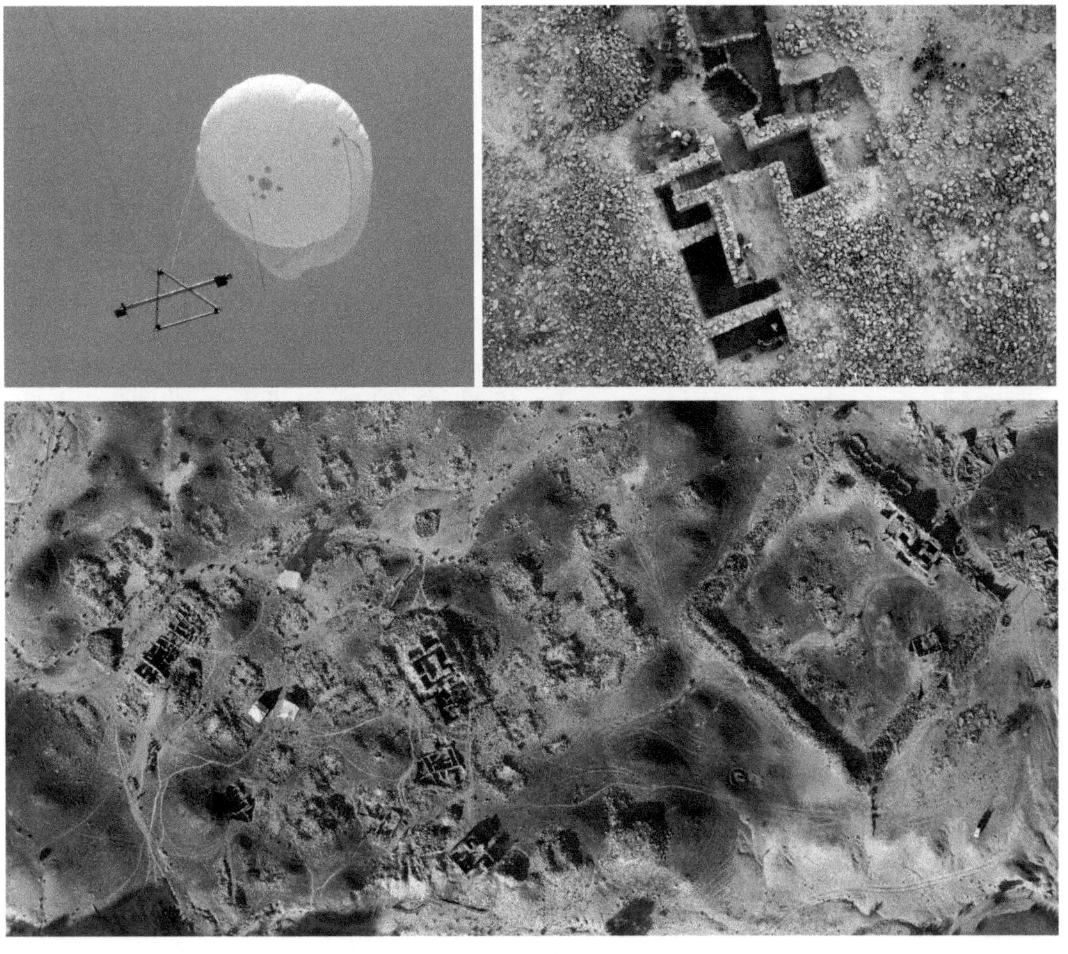

Figure 3. Top left: Helium balloon and photography platform in-flight; Top right: View of monumental Iron Age building at Khirbat en-Nahas (excavation area shown, ca. 15 x 25 m); Bottom: Image composite of KEN (ca. 10 hectares; square fortress = ca. 73 x 73 m). Shot from altitude of ca. 200 meters.

Figure 4. GigaPan 360° view of Khirbat en-Nahas, Jordan.
This imagery will be used in various 3D visualization environments.

The balloon was kept inflated throughout the 2-month long expedition season with little helium loss. It was sheltered each night in an on-site tent-like 'hanger'. In the 1930s, the Megiddo project similarly had to construct an on-site 'hanger' for their photography balloon – one of the earliest such applications (Guy 1932). OSDA 3.0 cameras were positioned on the balloon chasse system at the beginning of each day, allowing it to be ready for shooting within ten minutes.

Rapid recoverability of the balloon is possible by being tethered to the operator with a high strength, lightweight, *Spectra* line. This allows the operator to position the balloon over a specified area for shooting and recove it using the tether. Operation iss simple - the balloon is 'walked' until positioned over the area of interest. A live-feed from the air-borne cameras to a laptop on the ground allows a visual check to determine if the system is over the target.

Desert conditions required that the balloon system be r ugged, thus, there are few electronic components. The balloon itself was rugged and able to withstand UV radiation during its long deployment. Over 13,000 images were successfully captured amounting to over 200 Gigabytes of data during the 8-week season (Table 1). We are currently designing a more rugged aluminum platform, remote control to facilitate automated camera movement for both vertical and oblique photography, and stereo-video photography to enable 3D fly-through and fly-over of archaeological and cultural heritage sites.

b) GigaPan

GigaPan™ enabled us to use our Digital Single Lens Reflex (DSLR) cameras to produce multi-giga-pixel images at KEN not only to capture high-resolution panoramic views of the site (Figures 4,5) but also as a supplemental research tool. To enhance data acquisition and analysis, GigaPan photographs can be overlaid on the point clouds of archaeological features acquired digitally by LiDAR. Individual excavation areas were imaged using the GigaPan (Areas M, R, T, and W) and will then begeo-referenced with the billion-point LiDAR scans made with the LiDAR scanner.

The GigaPan imagery supplements the LiDAR data by providing more accurate, higher-resolution color information than delivered by LiDAR scanning. The process enhances the positional accuracy for the high-density three-dimensional LiDAR point clouds with color and texture information of comparable fidelity. The combined record is both photographically representative and three dimensionally precise down to the millimeter.

Figure 5. GigaPan view (180°) of the interior of a room in the Area W Iron Age building complex, KEN will used to supplement areas not covered by terrestrial LiDAR scans in the field.

c) LiDAR for Archaeological Field Work vs. Cultural Heritage Conservation

Recent developments in on-site scanning technologies add an important new field tool to the OSDA 3.0. Terrestrial LiDAR scanning was introduced in 2009 at KEN to augment recording, analysis and conservation efforts. The scanning was carried out with a Leica ScanStation 2™ that uses laser light to capture a collection of 3D points sampling the geometry and color of objects within its field of view. We acquired over 1.75 billion points in space comprising a high-resolution spatial record of the ancient fortress walls, gatehouse, residences and some 100 ancient unexcavated buildings visible on the site surface. The application of LiDAR scanning for documenting on-going excavations is relatively new: To date, most archaeology LiDAR applications have focused on recording sites as a means of ancient monument conservation (Barton 2009) and as a reconnaissance tool (McCoy and Ladefoged 2009). In the UK, relatively high-resolution LiDAR (ca. 1 – 2m) was used for site prospecting that produced results as good or better than aerial photography (Bewley, Crutchley, and Shell 2005). Our goal is to use LiDAR as a heuristic device to investigate archaeological data collected during the course of excavation as well as provide an accurate conservation record of a site.

i) LiDAR On-Site – Recording a 10 hectare archaeology site with sub-centimeter accuracy

The application in OSDA 3.0 application in Jordan exploited the full potential of sub-centimeter accuracy provided by LiDAR scanning for site recording and analyses. First, points were sampled every centimeter (compared to the 5-centimeter resolution of previous scans our team carried out in the Anza Borrego desert,

California, in preparation for the Jordan project – see Fox 2008a). LiDAR scanning at KEN produced ever-more-detailed scans, determining points approximately one millimeter apart. The LiDAR scans also yielded color and intensity information for each point, providing some insights about the properties of the material.
Successfully scanning a large archaeological site using terrestrial LiDAR technology poses a set of unique challenges, including proper scanner and target setup in and around sensitive archaeological artifacts dispersed over a site that in this case extends over 10 hectares. At KEN, line-of-sight distance from the scanner was limited, with many surfaces requiring scans from oblique angles adversely effecting the achievable range, speed and scan resolution. These challenges were further aggravated by electrical power requirements imposed by extended acquisition runs as well as environmental conditions such as dust and heat that impacted both the scanner and the control laptop. Finally, solid boundary conditions for high-precision geo-referencing of the acquired data had to be provided.

At least half a dozen LiDAR-based surveying devices using either time-of-flight or phase-based techniques (to determine the distance between the scanner and a target within line-of-sight of the scanner) are now commercially available. Most of these devices support a spherical scan envelope that is swept out one point at a time during the acquisition process. Most commonly, the laser is directed via a rotating or pivoting mirror, covering approximately 270° vertically (the remainder is commonly occluded by the system's tripod) and a rotating head allowing for 360° horizontal sweeps. Common sampling rates range between 15,000 and 150,000 points per second with a sample spacing that may be as small as 1mm for systems with an average laser spot size of approximately 4mm.

System Specification

The Leica ScanStation 2™ used at KEN is theoretically capable of sampling rates close to 50,000 points per second and acquisition densities of under 1mm^2 over distances of up-to 300 meters for objects with a 90% albedo. Under Jordan field conditions sample distances of 100 meters tend to be more realistic. The scanner weighs approximately 19 kg, the battery pack 12kg and its surveyor's tripod 9kg for a total of 40kg for the base system, excluding the acquisition laptop. When combined with a ruggedized transport case the overall weight more than doubles that requires an experienced operator and one assistant. The ScanStation2 has a maximal scan range of -45° to +45° vertically, requiring a two-pass sweep and 360° horizontally. User specific scan windows can be flexibly defined within this. In specifying a solid angle of interest, a physical resolution is derived using an average distance to target. With these two parameters in place, the physical resolution is computed in terms of samples per degree – in effect creating a variable spatial resolution based on the varying distance from scan surface to scanner. In the course of each scan, stationary surveyors' markers are scanned and enumerated as a spatial reference that can be used to align data collected from multiple scan positions. To drive the scanner, power is required as well as a standard *Ethernet* connection to a control terminal (limited by specification to ca. 100 meters). Electrically, the world-compatible power supply draws up to 400W for the scanner, although in practice the scanner averages 100W power, plus 60W average for the control laptop.

Field Deployment

A scanning sampling strategy is needed when tackling a large site as not all areas are of equal interest. The foremost concern when deploying the scanner is its placement relative to the object or panorama of interest. Uniformity of coverage is of utmost importance not all positions are visible from the scanner's head location can be represented resulting in voids in the final scan. Careful planning to cover a particular feature or surface from multiple angles is required to capture surface characteristics that would otherwise be represented as shadows with insufficient coverage. This involves a 'walk-through' with the excavation director to identify significant areas for detailed scanning.

The main excavated areas on the site were scanned at very high resolutions (1mm or 2mm). These high resolution scans have provided an impetus for developing new processing and visualization techniques capable of handling these large datasets. For key excavation areas, such as the fortress gatehouse, the following scanning goals were met:

1. Provide a clear and unobstructed view of features of interest.
2. Select locations where scanner is equidistant to features of most interest.
3. Provide multi-angulecoverage of non-planar surfaces – commonly, several scans were needed to avoid occlusions. In practice, this required changes of scanner elevation as well as position.
4. Solid geo-referencing support with limited interference from surveying targets (markers).
5. Limiting required scan distance to control errors introduced by variations in temperature and humidity.
6. Minimize the number of scan positions needed to achieve target objectives in respect to coverage and resolution.

In practice, handling ancient rooms with interesting features is relatively straightforward: the LiDAR technician can "walk" the scan lines, observe occlusions and make concessions of coverage between prospective scan positions. A digital camera and diligent analysis of occluded spatial regions also helps in planning. Constraints such as the suggested minimum distance from corners and surfaces helped to converge prospective scanner placement between multiple candidate sites.

To achieve greatest wide-area coverage of the site, a number of specialized scans directed exclusively at low-resolution panoramic coverage from an elevated position are best. Covering wide areas uniformly is best achieved by choosing a set of physically elevated candidate sites with the greatest visual coverage of the site. Co-locating these wide area scan sites with areas of interest For background scans, resolutions of 1cm at 50m are considered acceptable.

Empirically, the time flow for a scanner placement is estimated as follows:

- 5 minutes tripod deployment
- 5-10 minutes cabling and scanner set-up
- 8 minutes to boot up scanner
- 10 minutes to acquire background photography
- 5 minutes to acquire surveyor's markers
- 60 minutes for each 30 million points to be scanned (in practice)

Based on these parameters, it was important to minimize the number of scanner placements. For example, interior scans inside a building can be around 30 million points. Thus, the deployment cost noted here governs the time spent acquiring the scan. The only limitation for prospective scanner placement is direct line-of-site visibility to at least two separate surveyor's markers and those markers should be clearly visible and within 100 meters.

Geo-Referencing

Included in the scanning field deployment are four surveyor's markers. These markers are placed strategically to be seen from the majority of the scan locations, so that scan location can see at least two markers. The purpose of registering placement relative to the markers is so that several scans from different physical positions can be stitched together as one uniform scan. These stationary markers serve as the ideal reference points for chaining together a collection of scans.

With careful placement, four markers are sufficient for all but the most complicated and obscured areas of interest. Complications arise when multiple sites of interest are to be scanned and co-registered when one of two conditions is encountered:

1. Four target sites are insufficient to cover both areas of interest
2. The areas of interest are separated by more than 100 meters

In the course of the scanning campaign at KEN, both of these conditions occurred. To avoid the propagation of error likely to occur if targets were moved, they were never moved until all scanning was complete. Likewise, revisiting a site using a distant marker as a common registration link was ill-advised as error could easily be propagated in false or noisy measurements to the distant marker. To solve both problems, a strategy was established of systematically moving the scanner along the scan path in one direction through the site – thus containing error propagation by avoiding re-visitation of scanned areas.

Trying to manage the tradeoff and process of marker placement, scanner placement, context of scans and ordering of on-site scan work-flow proved to be formidable. In practice, the scan site and all regions of interest were thoroughly surveyed visually for an entire day before any equipment was deployed.

ii) Post-Excavation LiDAR Processing

The product of a scanning campaign is a collection of raw point clouds, one per scanner setup. These individual point clouds subsequently have to be merged, cleaned and geo-referenced (assigned latitude, longitude, and elevation), turning the data collection into an accurate representation of the field-site and excavation, suitable for visualization, analysis and co-registration with other digital data assets.

The scanning campaign at KEN yielded some 1.75 billion scanned points that were acquired following the scanning procedure outlined above. The process ensured that several GPS-referenced markers were visible in each cloud, establishing a common reference coordinate system for the entire collection and insured that scans could be accurately merged. Once merged, the overall data collection is cleaned to remove redundant, undesirable or extraneous points, for example, those of inadvertently scanned persons or equipment. Of particular importance is the treatment of points covering objects visible in multiple clouds, since not all objects are scanned equally well in every cloud: some are near the scanner, with high point density and precision, some are far away, or at an awkward angle with low density and precision. Consider a part of a wall that is visible in multiple clouds, with a differing scan quality in each. If the wall is covered well enough by points from a single 'best' cloud, the overall quality of the dataset can be improved by removing the lesser-quality wall points from the other clouds. The size of the resulting datasets have historically presented a major obstacle: just bringing the points to the screen quickly is a technical challenge. To address this, we have developed a novel system that allows billions of points to be interactively visualized on commodity hardware—however, the algorithms involved are not presented here.

Once the overall point cloud is available, a baseline record for the site is established that captures its spatial characteristics at a specific moment in time. With this record in place it is then possible to virtually explore the site, flexibly and freely, overcoming on-site limitations in respect to how easily particular locations may be accessible. When paired with intuitive interaction and display devices, visualization of this point data allows for the exploration of spatial relationships, correlations and measurement of the site and its artifacts (Figures 6-7). The use of room-sized display walls such as HIPerSpace, and virtual reality environments such as the NexCAVE and StarCAVE (e.g., Figures 21-22), provides a means to collaboratively study a vast amount of geo-referenced data of various types within an appropriate topographic context. At the same time, laptop-centric visualization provides a means to analyze data in the field, at a level of interactivity previously only seen in videogames. This may include analysis of the locations of artifacts, correlation of radiocarbon samples, and augmentation of close-up photographic records or references in the geo-referenced context provided by the point cloud.

We have also experimented with more structured analytical paradigms. One example is a grid system that partitions the site into fixed-size cells. The grid provides an organizational structure for annotations and measurements, and enables a constrained cell-by-cell navigation mode that complements the freeform 3D inspection interfaces. In addition to letting us view the site as a camera would—in perspective—the visualization system can also be used to cut out and draw arbitrary slabs and slices of the site orthographically, overlaid with a measurement grid and the real-world scale. This mode allows us to perform both qualitative and quantitative analyses. For example, we can quickly measure length visually by positioning and scaling the viewing slab appropriately. We can also easily obtain ground plans (Figure 8) or vertical sections (Figure 9) by positioning the viewing slab parallel or perpendicular to the ground, and selecting the appropriate slab thickness. Note that these types of analyses are made possible by the performance and interactivity of the underlying visualization system—the subject of interest is interactively manipulated into the desired position on the measurement grid, with continuous visual feedback, simply by moving through the virtual environment.

Figure 6 Viewed from above, a visualization of the LiDAR record resembles an aerial photograph. Note the differing levels of scan coverage corresponding to areas of greater and lesser interest: areas with minimal coverage appear as gaps or shadows between the more important regions scanned with a high point density. Shown here is the Iron Age fortress (ca. 73 x 73 m) in Jordan. Compare with Figure 3 (Image: CISA3/Calit2).

Figure 7. Viewing an area of interest up close reveals the detail acquired by the LiDAR scanning technique. A structure with interior rooms can be seen in full detail in the foreground, with sparser coverage of other areas and the background. Shown here is the gatehouse of the fortress at KEN (Image: CISA3/Calit2).

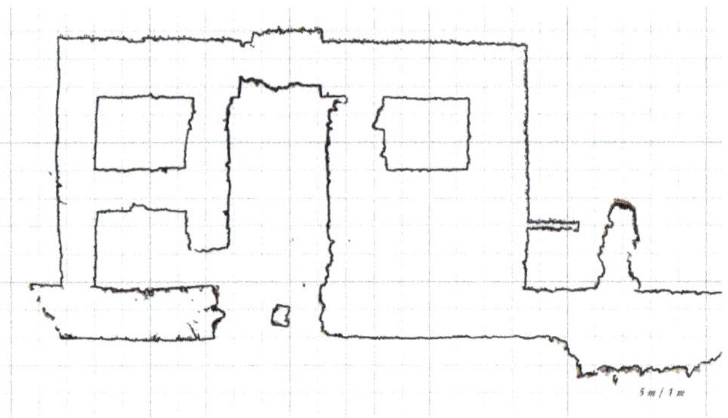

Figure 8. The point visualization tool can be used to present spatial relationships faithfully and to scale. The point cloud data is a precise record of the scanned area or objects, and can be used to perform measurements and other analyses. For example, we can obtain a floor plan of the Iron Age gatehouse with minimal effort by selecting a horizontal slice of points to view (Image: CISA3/Calit2).

Portable analytical tools – XRF in the field (EB-Y, TEL)

Portable high precision analytical tools are rapidly allowing researchers to bring the geoarchaeology laboratory to the field (Katz et al 2010). In recent years, the field of x-ray fluorescence (XRF) has been revolutionized with the development of portable devices. Instead of bringing samples to the laboratory, the researcher can now measure the bulk chemical composition of materials in the field and get results on the spot. Pioneering applications of this type of research have been carried out on early metallurgy sites in Israel (Yekutieli et al 2005; Vardi et al 2008) The portable XRF has various applications and more and more publications demonstrate the important role of integrated chemical analysis in field research. For example, archaeologists can measure and map soil composition (metalliferous pollution or other chemical variables of interest) spatially on site or along profiles in excavated sections, correctly identify objects already in the preliminary stage of research (bronze vs. copper vs. iron artifacts, gem stones, beads etc.), obtain typological 'chemical signatures' of artifacts as part of the sorting and cataloguing process in the field, etc. One of the fundamental advantages of portable device is the ability to measure materials that cannot be carried to the laboratory due to governmental rules or the large size of some cultural material.

The output information of the portable XRF is a complete list of chemical elements found in the sample, with their relative or absolute quantities (depending on instrument calibration and type of material). The measurement itself is non-destructive, rather quick (usually up to 300 seconds), and in most cases without the need of special sample preparation. The fast measurement process results in an exponential growth of digital data that should be organized and correctly linked to the sample information and excavation contexts. A major caveat of XRF application is that each measurement represents only a limited portion of the sample on the scale of a few millimeters (in diameter and in depth, depending on the specific structure of the device, its settings and the type

Figure 9. Alternatively, structures can be isolated and compared, such as the above common wall with obstructions removed. Note the automatic grid-line and scale overlay; the grid spacing and scale legend updates dynamically as the view is changed by the researcher. Here a north-face section has been made through the LiDAR data to illustrate a section illustrating two walls that delineate the doorways to two guard rooms (Image: CISA3/Calit2).

of material). This should be taken into account in the case of heterogeneous, corroded or patinated samples, and such metadata regarding sample characteristics must complement the XRF data.

At KEN, we integrated data from routine XRF measurements with the master GIS-based database of the excavation. More than 600 artifacts were measured with a portable *Bruker* XRF device, including soil samples, scarabs, copper ore, ceramics, slag and metal artifacts. The field measurements helped, for example, to identify high impurities of iron in the raw product of primary smelting and the lack of tin in most of the copper objects from the site; we also obtained elemental composition of hundreds of artifacts that had to stay back in Jordan. One such artifact was an arrowhead (Figure 10), that was found to be made of pure iron (now corroded, Figure 11). The XRF reading of this artifact (.PDZ file format in our system) is linked to its EDM number , and by it to its spatial coordinates, context (locus, basket), digital photography and other information included in the relevant *Access* database. The XRF database is currently being integrated into the visual analytics methodology described in this paper.

Figure 10. Arrowhead from Area W, a weapon made from iron, 2009 UCSD Jordan expedition (EDM# w09f2918; Photo: A. Gidding, UCSD Levantine Archaeology Lab).

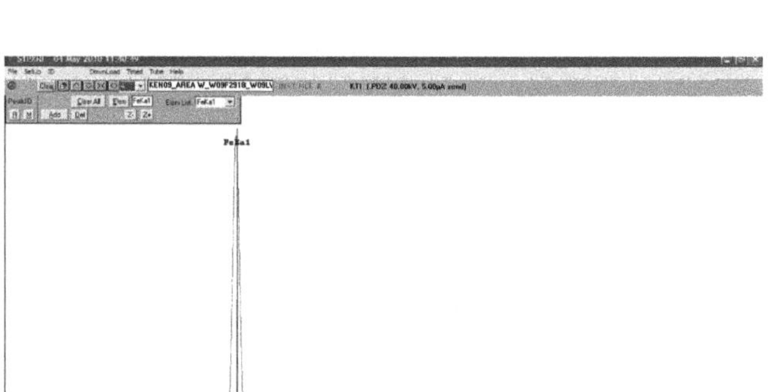

Fig.ure 11. XRF reading of the arrowhead (Figure 20) in initial examination mode showing that the artifact contains almost exclusively iron (Fe).

Portal Science, Cyber-Infrastructure and Cyber-Archaeology

a) MedArchNet – DAAHL

To share ancient settlement pattern and other archaeological data with as large an audience as possible, a cyber-infrastructure is needed that can promote the sharing and analyses of data in a communal manner. The NSF GEON "Cyberinfrastructure for the Geosciences" project (http://www.geongrid.org), whose Information Technology component is led by Chaitan Baru, serves as model for how the Earth Science community uses a cyberinfrastructure with "data portals" to facilitate delivery, discovery, access, and integration of distributed heterogeneous data sets (Baru in press). Our group has built a similar cyberinfrastructure to unify the many digital datasets and methods described in this paper. This 'portal science' application is called the *Mediterranean Archaeological Network* (*MedArchNet*, Figure 12) and is envisioned as a series of linked archaeological information or atlas nodes, each of which contains a regional database of archaeological sites that share a common database structure in order to facilitate rapid query and information retrieval and display within and across nodes in the network. To date, one digital archaeology atlas node is fully functional and utilized by hundreds of researchers. *MedArchNet* is a signature project of UCSD's CISA3/Calit2 and the Geo-Archaeological Information Applications (GAIA) Lab, Archaeological Research Institute at Arizona State University. The ultimate vision of *MedArchNet* is to develop a network of archaeological sites (from remote prehistory to the early 20th century – see http://medarchnet.org*MedArchNet* currently contains one active archaeological information nodes -the *Digital Archaeological Atlas of the Holy Land* (*DAAHL*) at http://daahl.ucsd.edu

The *MedArchNet* cyberinfrastructure provides secure and reliable storage of data from the field to the central data storage facility. It will provide authenticated, portal-based access to data, derived products, analysis, visualization, and GIS tools, collaboration spaces, etc, including provision of "publish/subscribe" interfaces for data, to enable a large user community to gain access to data and derived products. The cyberinfrastructure will manage heterogeneous archaeological data from a variety of sources, and support a community of contributors as well as users of the information, using a comprehensive authentication and authorization system to control access privileges of different classes of users.

The *MedArchNet* approach to archaeological site data envisions our data nodes as "switchboards" that contain top-level site and project metadata, plus bibliographic references and extensive use of linked resources outside the *MedArchNe*t data structure. It is not our goal to corral every bit of data about every site in the Mediterranean—an enterprise that would clearly be impossible even if it were desirable. Rather, the *MedArchNet* approach is designed to let researchers and the public easily find

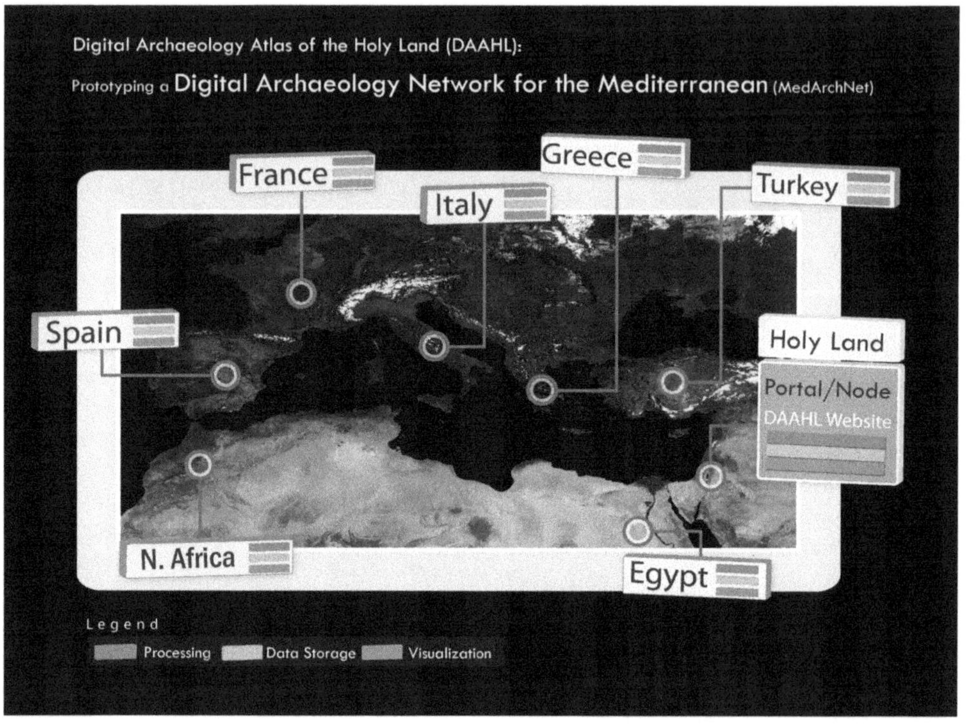

Figure 12. The MedArchNet website which highlights the most active node – Digital Archaeology Atlas of the Holy Land (http://medarchnet.calit2.net).

archaeological sites based on location and other attributes such as site type, features, time periods, etc., provide a mechanism for creating substantive maps linked to the various *MedArchNet* nodes, and then point the user to the locations of substantive research on the site, whether it be on- or off-line. The *MedArchNet* project serves to highlight the research of the archaeological community, rather than subsume it under the *MedArchNet* umbrella. Each data node maintains a table of data donors, including contact information and primary web sites, and each site contributed by a donor will be "branded" with the donor's information. Whenever a contributed site is displayed, the donor information is also shown, so the links to the donor's website are clearly shown, along with specific external resources for individual sites.

The *MedArchNet* project actively cooperates with research organizations and government agencies to develop new data nodes and applications. Each of our current nodes has received significant sponsorship. The *Digital Archaeological Atlas of the Holy Land* (DAAHL; Figure 13) is a sponsored project of the American Schools of Oriental Research, the flagship organization that coordinates North American archaeological research in the Levant (http://www.asor.org). As *MedArchNet* develops additional data nodes, we look forward to expanding our cooperative efforts with additional data donors, research organizations and government agencies. The *MedArchNet* databases are UTF-8 encoded, so they support multinational character sets; moreover, the Google Translation tool is included at the bottom of every *MedArchNet* web page, so the output can be translated into any available language at the touch of a button.

Figure 13. DAAHL's On-line Virtual Museum geo-references artifacts collected and recorded during excavations and displays them over a GoogleMaps platform. Shown here is a 9th c. BCE pottery sherd found at KEN hovering over a GoogleEarth image.

The DAAHLsite illustrates some of the content-rich methods it uses to disseminate data drawn from its database. Efforts to harvest archaeological site data from Israel and Palestine are currently underway. Another highly innovative feature is the *DAAHL's* Virtual Museum (Figure 13), which displays interactive, 3D objects at their original find locations through a Google Earth API—the user can manipulate the object in all three dimensions as well as the map itself.

MedArchNet is already having a significant research and education impact by providing easy online access to archaeological data and information. The *MedArchNet* hub and its data nodes are deployed to provide access to information contributed by each member of the *MedArchNet* "Virtual Organization". Via the hub, users are able to navigate back to the original member sites and databases to access the full information and related data from the respective site. Each data donor receives full recognition and credit for their contribution. Individual portals have been developed initially for ASOR (*DAAHL*) and other partnering groups. The network of linked portals support collaborations among users and provide a platform for initiating and sustaining discussions related to cross-site thematic areas of study.

In an era of rapidly expanding population and urban development, a system like *MedArchNet* can provide mechanisms to monitor archaeological site conditions over time and lessen the impact on cultural heritage resources by careful planning and significantly enhance site preservation and development potential in the Mediterranean basin. Furthermore, by uniting archaeological site metadata from many disparate datasets and organizations, the *MedArchNet* cyberinfrastructure will dramatically improve the ability of researchers to ask large-scale, cross-border questions of the archaeological data, providing fresh new insights into some of the most culturally meaningful regions on Earth.

b) 3D-Artifact Scanning

For much of its history, post-excavation archaeological research has relied upon manual drawings of artifacts that are often time-consuming and expensive to produce, subjective in detail and limited in their two-dimensional scope. Often, those drawings depict mere fragments of artifacts, making it difficult for archaeologists to recreate complete objects. To create the necessary 3-D algorithms that reflect the chronological and cultural 'address' of ancient potters in Jordan, our team uses NextEngine™ laser scanners to obtain triangulated meshes of the potsherds which can later be converted into any 3-D vector format (2-D images generate raster, or dot-based images, but vector, or shape-based images, are more amenable to mathematical manipulation). Since all the 3-D scans done by our team can also be imported into MATLAB (a numerical computing environment and programming language), we can compare and analyze, in the same format, both the 3-D scans and 2-D vectors of images taken from archaeological publications (see Foxb 2008).

The portable and relatively inexpensive NextEngine 3D scanners are"field operable" units.. They consist of two components - a turntable and a data capture device. The turntable allows for the object scanned to rotate 360° in front of the data capture device at specific intervals allowing for the scans to stitch together more accurately The data capture device consists of two primary components, a laser array and a digital 3mp camera in order to capture texture data. This combination allows us to create photorealistic models of the artifacts as portable as the computers that we used to scan the artifacts (Figures 14-15).

Figure 14. Undergraduate Caity Connolly uses NextEngine™ 3D scanner to image small Iron Age ceramic vessel in CISA3/Calit2 cyber-archaeology lab. This instrument was taken into the field to Jordan in fall, 2009 (Photo: Eric Jepsen, UCSD/Calit2).

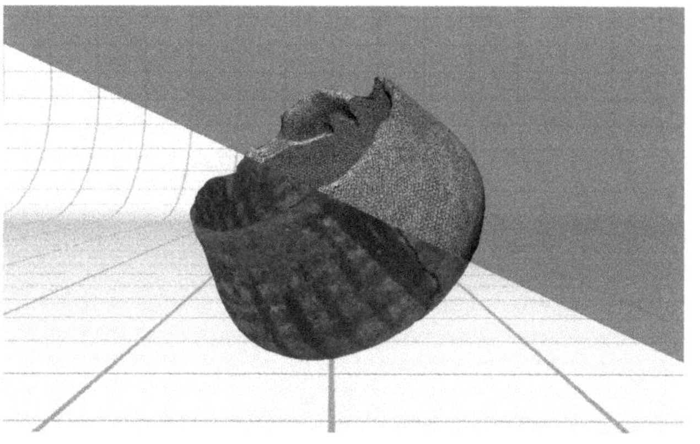

Figure 15. 3D scan of Iron Age ceramic cup from Area R, KEN. The 3D image file is used in the Pottery Informatics program described here (Image: CISA3/Calit2).

In 2009 we took the 3D scanners to the field for the first time. This wasan experiment to test how they would cope with the harsh desert conditions experienced within our field lab. In the CISA3 lab there is total climate, environment and light control that allow us to ensure that the artifacts are in the optimal conditions for high levels of scan accuracy (Guidi 2008). The problem in Jordan was how to adjust to far more challenging conditions while in the field. The ambient temperature was a particular problem, restricting us to be able to scan mostly in the early morning when the day was coolest. There was also the issue of the effect of ambient light affecting the accuracy of the scans. To solve this problem we simple built an inexpensive enclosure out of cardboard to create the most controlled environment possible in our field lab. With these minor adjustments we were able to achieve quality resultant scans on the field.

An area we are currently investigating is how to further streamline the 3D scanning process. One method to circumvent the unavoidable time expense required for 3D scanning was to purchase several scanners thus reducing scan time by a third. We are also trying to automate a number of the processes so that less student time is required to scan and process these data. In general, now that we have begun 3D scanning of artifacts, the NextEngine scanner has become a standard tool in our archaeological 'digital tool box.'

c) Artifact Informatics: Pottery Informatics Queryable Database – PIQD

Beginning in 2008 we began to develop an informatics database for the processing of diagnostic Iron Age (ca. 1200 – 500 BCE) pottery sherds. The goal of this project was to develop a method of digitizing artifacts collected in the field within an analytical framework similar to that used in bioinformatics for the analysis of protein and DNA sequences (Altschul et al. 1990; see below). The increased precision achieved through these analyses exposed a number of drawbacks to the use of only digitized 2D illustrations of ceramics. One of the most prevalent problems being the relative subjectivity of the professional illustrators attempt to draw a 2D profile representative of the whole sherd, its stance, and measured diameter. In order to achieve the most accurate representation of the diagnostic pottery sherds recovered from excavation, we started in this season to scan pottery using the NextEngine™ 3D scanner (see Figures 14-15), which allows us to confidently attain accuracy within 0.12 mm for each sherd scanned. Once the sherd is scanned it is imported in Matlab where its proper stance, rim diameter, and profile are extracted at this same level of precision. In addition, we used the 3D scanner to scan other artifacts that are even more geometrically complex.

Although there has been increasing interest among archaeological projects aimed at digitizing various aspects of their archaeological datasets, there has been no satisfactory solution for integrating these different projects' data for cross-regional comparison and analyses. Most researchers still must rely upon printed publication reports to conduct any form of regional study. In general, this medium is very limited in its ability to inform the reader of the nuances of the material culture collected at the site, such as architecture, stratigraphy, ceramic, and other artifact assemblages.

In order to circumvent a number of these drawbacks, a comprehensive online queryable digital database called the Pottery Informatics Query Database (PIQD) was started for the Iron Age ceramic assemblages of the southern Levant (http://daahl.ucsd.edu/PIQD/PotteryInformatics.php). This project was begun in 2008 between Smith and Levy at UCSD CISA3/Calit2 and Avshalom Karasik and Uzy Smilansky at the Weizmann Institute of Science-Hebrew University in Israel. It is a new online tool designed to enable researchers to test their own interpretations and models against the ever-expanding digital medium of ceramic datasets in ways that conventional print data cannot provide. Where the PIQD differs from other online archaeological databases that may archive published 2D vectorized images or 3D models of ceramics, is in query-ability – in particular, with objective mathematically based algorithms of artifact (ceramic) profiles. This project uses recent technological advances developed by Karasik and

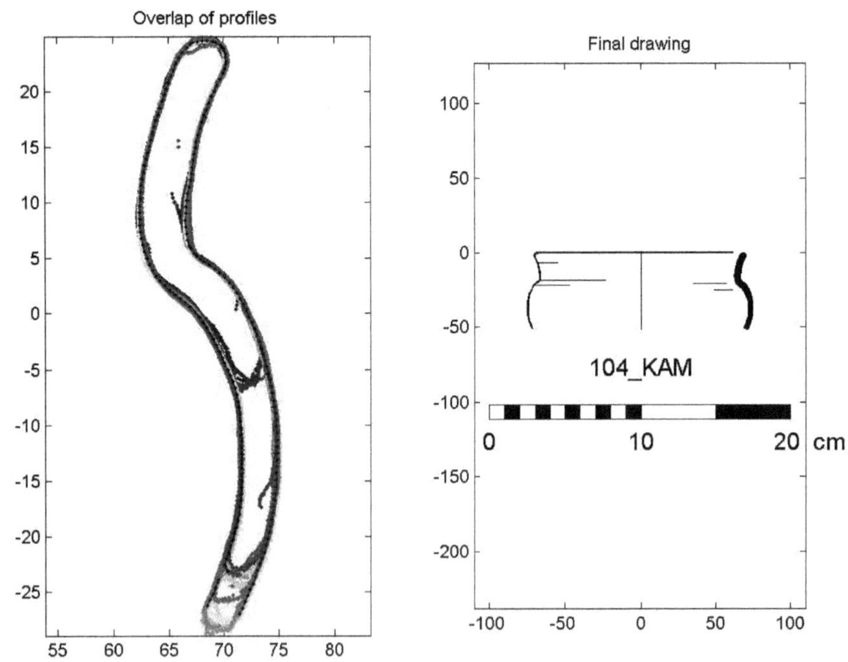

Figure 16. Mathematically extracted 2D profile of 3D scanned bowl from the Iron Age site of Khirbat am-Malayqtah, Jordan and creation of a pottery drawing suitable for publication .

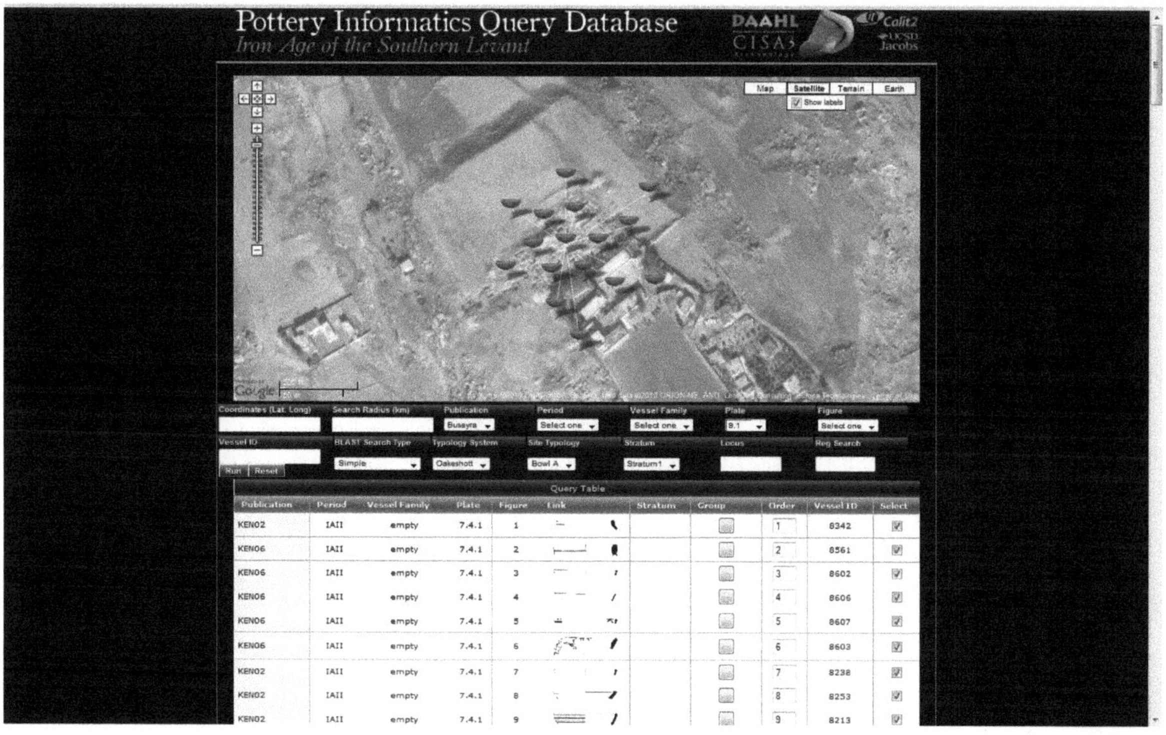

Figure 17. The Pottery Informatics Query Database GUI Page

Smilansky (Karasik 2008) to mathematically encode and store the ceramic profile data as complex algorithms (Figure 16). Three mathematical representation functions (radius, tangent and curvature) measure various scales of differences in vessel form, stance, and rim diameter, which can be used to determine in an objective manner the statistical difference between ceramic shapes. This technique combined with several methods of cluster and discriminate analysis has been used to construct objective mathematically based typologies and ceramic prototypes. Specifically for the PIQD, the three functions enable the rapid search of a whole database of digitally stored vessels in an objective mathematically grounded approach. In this sense, these queries are similar to online BLAST searches (Altschul et al. 1990) developed in the field of genetics in being able to rapidly associate large quantities of digital vessel profiles to each other based on similar morphological traits.

The PIQD has been designed using MySQL, PHP, Ajax, and Javascript with an imbedded Google Maps API to provide a fully queryable spatial environment for the user. In essence, the PIQD is an open source GIS. GoogleEarth and Google Maps function as a real-time spatial display for all the ceramics' coordinate information. MySQL functions as the server database to organize all the stored ceramics profiles, curvature functions, 3D scans, and metadata, while PHP and Javascript are used to query the database.

Currently the PIQD is being designed to run autonomously in its incorporation of new vessel profiles and the computation of its mathematically based typology query system. Users are able to directly upload their raw data and have the PIQD properly store it, insert the metadata into tables, and run the needed cluster analyses. A set of automatic error checking mechanisms were developed to insure that only error-free data are accepted into the PIQD for study. By this means, the PIQD is able to expand exponentially as multiple researchers can contribute their ceramic assemblage data to the overall PIQD without the need for a technician to oversee every new entry. At present, there are over 10,000 ceramic figures and their associated metadata from Iron Age Edom that have been incorporated into the database (Figure 17). The immediate goal is to achieve complete coverage of the Iron Age for all of the Southern Levant, which will reach into the 100,000's.

Finally, two daughter programs were also developed to further facilitate the original contributions of archaeologists and their publication. These programs plug directly into the PIQD. First, the *PlateMaker* program enables publication quality plates of queried ceramics to be auto-generated and manipulated on the fly (Figure 18). Tables are dynamically linked to the displayed figures on the plate so that either the reordering of the table or the plate always remains synced. Second, the *MasterTable* is a dynamic spreadsheet that can expand or contract hundreds of fields and rows of data stored within the PIQD based on a simple user interface. It can immediately update changes made to the data through direct user input or barcode scanners. Finally, a series of Archaeological functions were developed to further facilitate common but complex manipulations of archaeological data sets. These functions can be accessed for any future daughter programs related to archaeology.

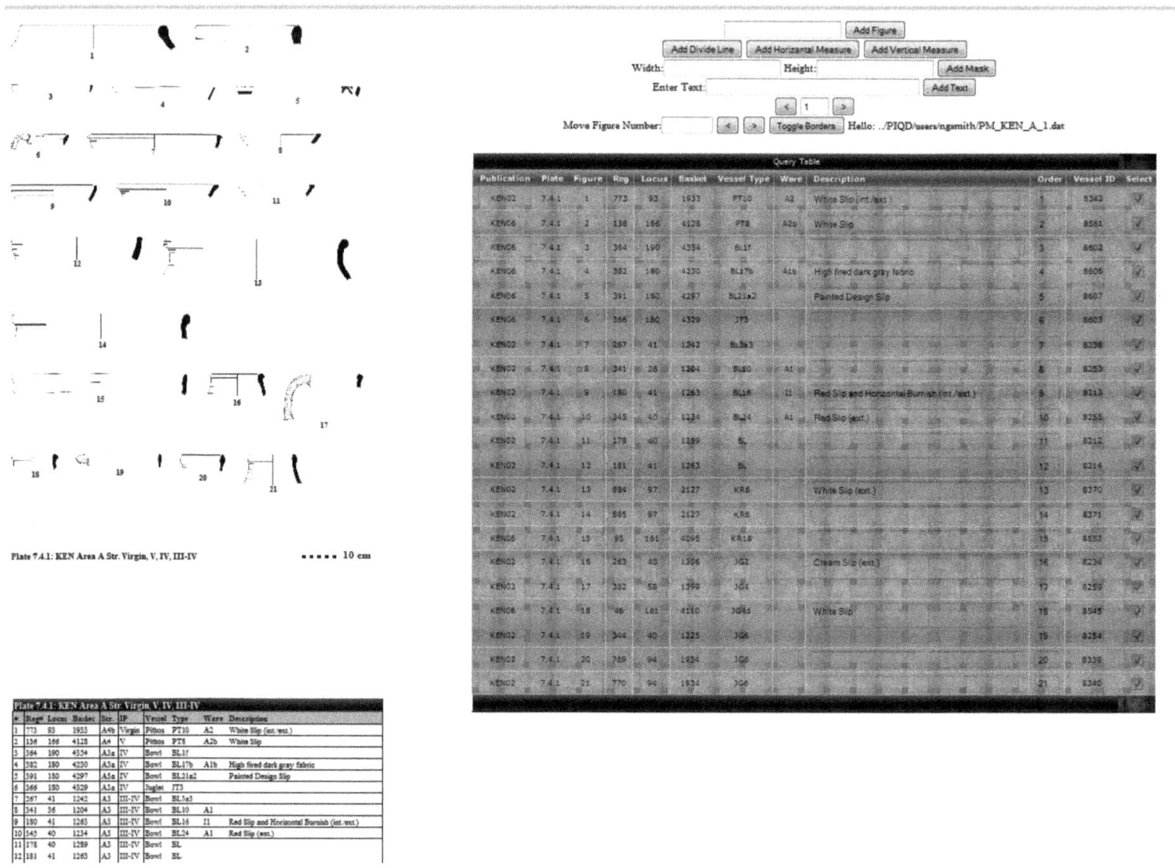

Figure 18. The *PlateMaker*: A daughter program of the PIQD designed for the rapid deployment of publication quality ceramic plates

The PIQD is not limited to only the Iron Age but can be adapted for any ceramic period. For example, we are also developing a mirror of the Iron Age pottery informatics database for the Early Bronze Age in the Southern Levant. In a period such as the Early Bronze Age, where variation in the pottery is not especially significant through time (Dever 1973), this digital technology can provide an important new way to make inferences about the material within the site itself. The PIQD has the potential for revolutionizing how ceramic assemblages are analyzed, typologies constructed, and how regional analyses are conducted. As the PIQD grows the dependence on the printed medium will subside as equivalent data can be found on the PIQD but by a much more rapid and accessible manner. The PIQD, itself being an online tool part of the MedArchNet, means that it can be utilized for data analyses not only in the Southern Levant but wherever a portal has been established. Moreover, the PIQD is not limited by the Iron Age period or ceramics but was designed to eventually be adapted for various forms of material culture and archaeological periods.

Toward a Cyber-Archaeology: Data acquisition techniques and visualization:

Multi-spectral imaging holds great promise for the transformation of traditional dirt archaeology into its digitally-enabled form. Visual records in the form of static images and videos in the visual range provide a baseline, capturing the spatial and temporal characteristics of a site. These can be further augmented with thermal images capturing sub-surface characteristics, photogrammetry techniques associating physical dimensions with image data, 3D topological scans yielding high-resolution point cloud collections via light detection and ranging (LiDAR) or stereo photography techniques, all of which are backed up by "traditional" on-site surveying techniques. While it is still possible to use old fashion dumpy-levels and measuring tapes, the trend is for research oriented projects to employ some form of on-site digital archaeology recording of the excavation process (Bunimovitz and Lederman 2009, Daly and Evans 2005, Levy 2010, Levy and Smith 2007). The compelling reasons for the transition to Digital Enabled Archaeology are accompanied by a set of daunting challenges associated with the exponential growth of data that has to be properly recorded, processed, fused, analyzed, archived and preserved. Some of the required workflows, visualization and analysis techniques as well as underlying infrastructure are extensive and illustrated here.

Visual Analytics and Instruments

a) HIPerSpace – A Visual Analytics Cyber-Collaboratory

The complexity and amount of data that we are confronted with as a result of the digitally enabled archeology paradigm, creates unique challenges for the accurate and efficient analysis of data in formats appropriate for the tasks at hand. Data records acquired with the acquisition tools described here are massive in size and multidimensional in nature. Moreover, comprehensive data analysis usually requires simultaneous access to multiple data sources represented in domain specific as well as synthesized formats, requiring an environment that co-locates domain specialists and data assets while enabling interactive, visual data analysis and reasoning.

To address this challenge, our team has developed the concept of visualization portals, scalable, high-resolution tiled display environments, operating at tens to hundreds of megapixels resolution, while enabling intuitive, information rich and rapid visual analytics, capitalizing on multiple human senses to convey higher-dimensional content. The "gold-standard" for these environments has been HIPerSpace, a 1/3 gigapixel resolution visualization environment, with the ability to collocate vast data collections in real-time, for a broad set of 2D and 3D formats (http://hiperspace.calit2.net). HIPerSpace is powered by a scalable and hardware agnostic visualization middleware called CGLX (Doerr and Kuester 2010), which was designed to provide a common visualization platform, supporting networked, scalable, multi-tile 2D and 3D visualization environments. With this infrastructure in place, it has been possible to also transform the traditional data analysis workspace into its digital equivalent, where data in the form of images, videos, 3D models, publications, web references, simulations, etc. can be co-located in one room-sized collaborative digital workspace. Figure 19shows the co-location of multiple different artifacts from UCSD excavations in Jordan that can be manually or automatically sorted, clustered, segmented and filtered, effectively providing immediate control over how data is being represented and explored. Figures 19 and 20 show and interactive visualization of a large-scale point cloud data set obtained via LiDAR scanning, which provides an accurate 3D view of a field site and means to explore the site, topological artifacts and spatial relationships. HIPerSpace and other environments similar to it, described below, are a portal into vast archeological data collections and allow research team to harness both data and even more importantly the human assets, the researchers on the forefront of new discoveries.

Figure 19. Interactive visualization and co-location of different artifacts from Jordan on HIPerSpace. The HIPerSpace is made with 72 'off-the-shelf' Dell 3007WFP-HC, 30" displays or tiles (Photo: Falko Kuester, Calit2) .

Figure 20. Interactive visualization of large-scale HD composite aerial photograph, providing an georeferenced view of KEN, and means to explore the site, topological artifacts and spatial relationships (Photo: Falko Kuester, Calit2) .

The HIPerSpace has proven to be a critical element for rapid visual comparison of theory with massive experimental data collections, enabling transdisciplinary teams of scientists to swiftly validate and comprehend theory and practice (Figure 19-20). With visualization as a unifying language anchored in mathematics and physics, OptIPortals (http://wiki.optiputer.net/optiportal/index.php/Main_Page ; DeFanti 2008b) provide a unique mechanism to communicate information in a universal format that allows hard domain problems to be approached by co-located or spatially separated interdisciplinary research teams.

The StarCAVE

Archaeological data is especially suited for building scientific visualization paradigms and creating virtual environments for Cyber Archeology. At Calit2, our group has focused on four main approaches to help create cyber-infrastructures for archaeology (Knabb, Schultze, and Levy in press). The areas of archaeological visualization research center on use of the 3D immersive visualization environment called the "StarCAVE" (Defanti et al 2008a, Levy et al 2008 [Figure 21].:

1. The StarCAVE is used to display compelling visual imagery using data collected in the fieldincludeing spatial data, digital images, and site reports. This is an important and efficient way to disseminate complex information in a manner that is straightforward and comprehensible (Forte and Siliotti 1997). Furthermore, researchers are able to investigate each excavation area or unit and the data collected from these units in three-dimensions.
2. After visualizing an archaeological site in the StarCAVE it is used as a heuristic tool. One is able to revisit the site and data again and again without ever going back to the field. In this manner we can

investigate the topological and spatial relationship between artifacts, features and other areas of the site and how these changed through time.

3. The StarCAVE is also used as a virtual reality Geographic Information System. GIS programs, such as ESRI's *ArcView*, allow the user to browse, query and manipulate the database, but these cannot handle three-dimensional data. Virtual reality technology such as the StarCAVE does, however, operate in a three-dimensional virtual world, and is sophisticated enough to simulate a 'real' environment. The precise location of each recorded artifact, feature and locus is, in a manner of speaking, put back together again. Additionally, the incorporation of GIS databases into the virtual reality model imparts the archaeologist with the ability to perform spatial and statistical analyses similar to the tools available in standard GIS programs.
4. The StarCAVE contributes to cultural heritage preservation. The unfortunate consequence of archaeological research is we destroy that which we study. Now, incorporating the many sources of data we collect into a virtual reconstruction of the site preserves a record of what was destroyed during the course of excavation in a manner that is more compelling to a large audience than a two-dimensional representation.

Our 3D virtual reality visualization application is designed to run within COVISE. COVISE allows us to design software applications at the desktop and then run them in a large variety of virtual environments, including Power Walls, CAVEs, and tiled display walls, including the StarCAVE at Calit2. The StarCAVE is a 5-walled, rear-projected, 360 degrees virtual reality device. It uses 34 high definition (1080p) projectors to generate passive stereo images on 15 screens and the floor. We use head tracking and a 3D input device to navigate and interact with virtual environments. Our application uses COVISE's VRML loader, as well as the previously described artifacts plugin. We converted the Google Sketchup model to a VRML file in order to load it directly into COVISE. Once that happened, we were able to navigate around the terrain and to all the excavation sites modeled, as well as get a bird's eye view of the entire area. In addition to the navigation around the scene, the users can scale the size of the excavation area so that they can display the 3D structures life size, which makes them appear as if the users were on site. Alternatively, the size can be scaled down so that the area looks like a small model where everything is within arm's reach. The ability to change the scale helps point to specific locations, and will help in the artifact display mode to select larger volumes of artifacts than the person can comfortably reach when displayed life size. However, the ability of the display to convey size like in the real world allows the user to not only perceive objects at their original level of scale, but also to measure distances and sizes with their hands, or a virtual measuring tape.

The current implementation allows the user to optionally display or hide the artifacts or the Sketchup model, so that one can focus on either without cluttering the screen with the other. This functionality is selected directly from within the virtual environment using a 3D menu. The 3D menu API we use is part of the COVISE framework and allows the programmer to add buttons, check boxes, sliders, dials, submenus, and custom dialog windows. Although the menus are mostly 2D and resemble menu systems in desktop systems, they can be moved around

Figure 21. Jürgen Schultz and Kyle Knabb demonstrate Iron Age building and excavation section through associated slag mound from KEN in the StarCAVE. The StarCAVE is a 360° total immersive 3D VR environment with additional imagery on the floor (Photo: Eric Jepsen, UCSD-Cali2).

freely in the 3D environment so that they are not in the way when exploring the virtual world.

NexCAVE

Calit2 virtual reality researchers offer archaeologists a 3-to-21-panel, 3-D visualization display made from newly available synchronized 3D HDTVs. The technology, dubbed "NexCAVE," was designed and developed by Calit2 Research Scientists.. The NexCAVE technology was developed at the behest of Saudi Arabia's King Abdullah University of Science and Technology (KAUST), which established a special partnership with UC San Diego last year to collaborate on world-class visualization and virtual-reality research and training activities (Figure 22).

When paired with polarized stereoscopic glasses, the NexCAVE's modular, micropolarized panels and related software ill make it possible for archaeologists to visualize massive datasets in three dimensions, at unprecedented speeds and at a level of detail impossible to obtain on a typical computer display. The NexCAVE's technology delivers a faithful, deep 3-D experience with great color saturation, contrast and very good stereo separation. The JVC panels' xpol technology circularly polarizes successive lines of the screen clockwise and anticlockwise and the glasses you wear make you see, in each eye, either the clockwise or anticlockwise images. This way, the data appears in three dimensions. Since these HDTVs are very bright and high-contrast, 3-D data in motion can be viewed in a normally-lit environment, even with the lights in the room fully on.

The NexCAVE's data resolution is superb, close to human visual acuity (or 20/20 vision). The 10-panel, 3-column NexCAVE has a ~6000x1500 pixel resolution, while the 21-panel, 7-column version built for KAUST has a ~15,000x1500-pixel resolution. The NexCAVE's LCD screens are scalloped "like turtle shells," which allows the screens' bezels (frames) to be minimized because the screens are tucked behind one another. This works well in 3-D because the virtual reality illusion is so strong that you don't even see the screens and bezels as "windows," just the 3-D images in motion and stereo.

NexCAVE's specially designed COVISE software (developed at Germany's University of Stuttgart) and CGLX software (developed at UCSD) combine the latest developments from the world of real-time graphics and PC hardware with high-end Nvidia game engines. The Calit2 and KAUST NexCAVEs are connected via 10 gigabit/second networks, which allows researchers at KAUST to collaborate remotely with UCSD colleagues. NexCAVEs are being designed and built for several new partners around the world. For the opening of the KAUST celebration in September, 2009, archaeological data from the UCSD Levantine Archaeology Lab excavations in Jordan that had been originally modeled for the StarCAVE was featured in the NexCAVE. We are now building a portable NexCAVE prototype that will help answer the problem of 'data avalanche' now facing field archaeologists.

Figure 22. The 21-Panel Xpol LCD stereo NexCAVE with UCSD – Department of Antiquities of Jordan excavation data from KEN on display at the opening of the King Abdullah University of Science and Technology (KAUST), September, 2009. Shown here are Tom Levy, UCSD (left) and Sami Almagouth, KAUST, with the NexCAVE demo in Saudi Arabia (Photo: T. DeFanti, Calit2).

Conclusion – the data avalanche

This paper has presented an integrated system of on-site digital archaeology coupled with an active cyberinfrastructure for Levantine archaeology with data that is capable of being viewed with a variety of 3D visual instruments. The discussion of data acquisition techniques revolves around the problem of how to acquire digital data in the first place. We refer to this as creating a 'digitally enabled archaeology.' Here the type of data and how it can be obtained is outlined in relation to a number of imaging techniques used by the UCSD CISA3/Calit2 team. These include Total Station, GPS, LiDAR, balloon based airborne imaging, and 3D artifact scanning. How these data are synthesized is also discussed. To achieve a truly "Cyber-Archaeology" the global networks that portal science cyberinfrastructures such as the Mediterranean Archaeology Network (MedArchNet) and its Pottery Informatics Queryable Database are described. Finally, a number of visualization paradigms and environments for Cyber-Archaeology that take advantage of these datasets are also described. The ones used by the CISA3/Calit2 team include the HiPerSpace, StarCAVE, and NexCAVE.

Based on the Jordan field work discussed here, our team has experienced a 'data avalanche.' This is highlighted in Table 1 that outlines the range of digital data collection instruments used in the field in 2007 versus those used in 2009. The exponential increase in digital data (from 172 GB in 2007 to 1,373 GB in 2009 is astounding. How will archaeologists deal with this data avalanche in their field projects in the future? Our team is working on a portable NexCAVE that will have software, computing power and portability needed to thread together all these rich sources of data in a manageable manner. One of our goals is to equip our helium balloon platform with stereo digital video cameras to allow our x,y and z archaeology data to be fully integrated into the 3D visualization instruments currently evolving.

Data Collection Instrument	2007 Excavation Season	2009 Excavation Season
3D NextEngine Scanner (Artifacts)	0	168 GB
High Res Digital Artifact Photography	70 GB	253 GB
High Res Digital Site Photography	10 GB	52 GB
Balloon-based Stereo Digital Photography	0	253 GB
Gigapan Panorama Photography	0	15 GB
Terrestrial LIDAR Scanning	0	500 GB
GIS Data	20 GB	30 GB
HD Video	72 GB	120 GB
TOTAL	172 GB	1,373 GB

Table 1. Data avalanche from 2007 to 2009, UCSD CISA3/Calit2 expeditions in Jordan.

Acknowledgements

Much of this project was funded by two UCSD Chancellor Collaboratory grants awarded for the 2008/2009 and 2009/2010 academic years and the CISA3/Calit2 Cyber-Archaeology Laboratory at UCSD. Thanks especially to Ramesh Rao, Director, and the staff at Calit2, San Diego Division for support of the balloon project. Thanks also to Dr. Fawwaz Al-Khraysheh, Director General, Department of Antiquities of Jordan for hi support.

References

Altschul, S. F., W. Gish, M. Webb, E. W. Myers, and D. J. Lipman. 1990. Basic Local Alignment Search Tool. *Jorunal of Molecular Biology* 215:403-410.

Barton, J. 2009. 3D laser scanning and the conservation of earthen architecture: a case study at the UNESCO World Heritage Site Merv, Turkmenistan. *World Archaeology* 41:489 - 504.

Baru, C. in press. "Portal Science Today: Building a Web Portal System for Archaeology," in *Portal Science and Archaeology - Views from the Mediterranean Lands*. Edited by T. E. Levy, S. H. Savage, C. Baru, and O. LaBianca. London: Equinox.

Bewley, R. H., S. P. Crutchley, and C. A. Shell. 2005. New light on an ancient landscape: Lidar survey in the Stonehenge world heritage site. *Antiquity*. 79:636-647.

Bunimovitz, S., and Z. Lederman. 2009. The Archaeology of Broder Communities - Renewed Excavations at Tel Beth-Shemesh, Part 1: The Iron Age. *Near Eastern Archaeology*. 72:114-142.

Daly, P., and T. L. Evans. Editors. 2005. *Digital Archaeology - Bridging Method and Theory*. London: Routledge.

DeFanti, T., Dawe, G., Sandin, D., Schulze, J., Otto, P., Girado, J., Kuester, F., Smarr, L., and Rao, R., 2008a "The StarCave, a third-generation cave and virtual reality Optiportal," *Future Generation Computer Systems/The International Journal of Grid Computing: Theory, Methods and Applications, Elsevier B.V., Vol 25, Issue 2, Nov. 2008.* Also: http://ivl.calit2.net/wiki/index.php/StarCAVE_Measurement_Diagrams.

DeFanti, T., Leigh, J., Renambot, L., Jeong, B., Verlo, A, Long, L., Brown, M., Sandin, D., Vishwanath, V., Liu, Q., Katz, M., Papadopoulos, P., Keefe, J., Hidley, G., Dawe, G., Kaufman, I., Glogowski, B., Doerr, K.-U., Girado, J., Schulze, J., Kuester, F., Smarr, L., 2008b "The OptIPortal, a Scalable Visualization, Storage, and Computing Interface Device for the OptIPuter," *Future Generation Computer Systems/The International Journal of Grid Computing: Theory, Methods and Applications*, Elsevier B.V., Vol 25, Issue 2, Nov. 2008.

Doerr, K. U., and F. Kuester. 2010. "CGLX: A Scalable, High-performance Visualization Framework for Networked Display Environments,". *IEEE Transactions on Visualization and Computer Graphics* 99.

Eiteljorg II, H. 1994. "Using a Total Station," in *CSA Newsletter*, vol. VII.

Forte, M., and A. Siliotti. Editors. 1997. *Virtual Archaeology: Re-creating Ancient Worlds.* New York Harry N. Abrams.

Forte, M. 2008. *Virtual archaeology: communication in 3D and ecological thinking*. BAR International Series 1805:21-35. Oxford: BAR Publishing.

Fox, T. 2008a. "CISA3 Researchers Analyze Native American Site in Push for Digital Archaeology," in *Calit2 Newsroom*, vol. 2008, pp. http://www.calit2.net/newsroom/article.php?id=1284.

Fox, T. 2008b. "CISA3 Archaeologists Develop 3-D Images of Artifacts for Digital Pottery Informatics Database," in *California Institute for Telecommunications and Information Technology Newsroom*, vol. 2010. San Diego: UCSD Calit2.Access date: May 7; http://www.calit2.net/newsroom/article.php?id=1357

Guidi, G., F. Remondino, G. Morlando, A. Del Mastio, F. Uccheddu, and A. Pelagotti. 2008. "Preformances evaluation of a low cost active sensor for cultural heritage documentation."

Guy, P.L.O. 1932 Balloon Photography and Archaeological Excavation. *Antiquity* 6:148-155

Haiman, M. 1996. Early Bronze Age IV Settlement Pattern of the Negev and Sinai Deserts: View from Small Marginal Temporary Sites. *Bulletin of the American Schools of Oriental Research* 303:1 - 32.

Karasik, A. 2008. "Applications of 3D technology as a

research tool in archaeological ceramic analysis. ," in *Beyond Illustration: 2D and 3D Digital Technology as Tools for Discovery in Archaeology.* Edited by Bernard Frischer and Anastasia Dakouri-Hild. pp. 103-116. British Archaeological Reports International Series 1805. Oxford: BAR Publishing.

Katz, O., Cabanes, D., Weiner, S., Maeir, A. M., Boaretto, E., and Shahack-Gross, R. 2010. Rapid Phytolith Extraction for on-Site Analysis of Phytolith Concentrations and Assemblages: An Application at Tell es-Safi/Gath, Israel. Journal of Archaeological Science 37(7): 1557–63.

Knabb, K., J. Schultze, and T. E. Levy. in press. "StarCAVE: 3D Virtual Reality Environments for Archaeology" in *Portal Science and Archaeology – Views from the Mediterranean Lands*. Edited by T. E. Levy, S. Savage, C. Baru, and O. LaBianca, pp. 16 pages. London: Equinox.

Levy, T. E. Editor. 1987. *Shiqmim I: Studies Concerning Chalcolithic Societies in the Northern Negev Desert, Israel (1982-1984)*. Oxford: BAR International Series 356.

—. 2010. "The New Pragmatism: Integrating Anthropological, Digital, and Historical Biblical Archaeologies," in *Historical Biblical Archaeology and the Future – The New Pragmatism*. Edited by T. E. Levy, pp. 3-44. London: Equinox.

Levy, T. E., J. D. Anderson, M. Waggoner, N. Smith, A. Muniz, and R. B. Adams. 2001. Interface: Archaeology and Technology - Digital Archaeology 2001: GIS-Based Excavation Recording in Jordan. *The SAA Archaeological Record* 1:23 - 29.

Levy, T. E., T. Higham, C. Bronk Ramsey, N. G. Smith, E. Ben-Yosef, M. Robinson, S. Münger, K. Knabb, J. Schulze, P., M. Najjar, and L. Tauxe. 2008. High-precision radiocarbon dating and historical biblical archaeology in southern Jordan. *Proceedings of the National Academy of Sciences* 105:16460-16465.

Levy, T. E., and N. G. Smith. 2007. "On-Site Digital Archaeology: GIS-Based Excavation Recording in Southern Jordan," in *Crossing Jordan – North American Contributions to the Archaeology of Jordan*. Edited by T. E. Levy, M. Daviau, R. Younker, and M. M. Shaer, pp. 47-58. London: Equinox.

McCoy, M., and T. Ladefoged. 2009. New Developments in the Use of Spatial Technology in Archaeology. *Journal of Archaeological Research* 17:263-295.

Vardi, J., S. Shilstein, S. Shalev, and Y. Yekutieli. 2008. The Early Bronze Age IV Chipped and Ground Stone Assemblage of 'En Yahav and its Relation to Copper Smelting Activities. *Journal of the Israel Prehistoric Society* 38:1-20.

Yekutieli, Y., S. Shalev, and S. Shilstein. 2005. 'En Yahav - A Copper Smelting Site in the 'Arava. *Bulletin American School of Oriental Research* 340:35-55.

www.ingramcontent.com/pod-product-compliance
Lightning Source LLC
LaVergne TN
LVHW071216210926
848482LV00015B/824